Rheology of Drag Reducing Fluids

Aroon Shenoy

Rheology of Drag Reducing Fluids

Springer

Aroon Shenoy
Waterford Hills
Germantown, MD, USA

ISBN 978-3-030-40044-6 ISBN 978-3-030-40045-3 (eBook)
https://doi.org/10.1007/978-3-030-40045-3

This Springer imprint is published by the registered company Springer Nature Switzerland AG
The registered company address is: Gewerbestrasse 11, 6330 Cham, Switzerland

This book is dedicated to my dear wife,
Jyoti Shenoy
(nee Prasanna Nagesh Prabhu),
who has been a constant source of inspiration
throughout my research career.

Preface

The idea that very minute quantities of certain specific materials create a dramatic drag reduction effect in turbulent flow of fluids is highly intriguing, and this led several researchers to tread the path to discover and explore this phenomenon over the years. There has been a realization that the type of material that can demonstrate drag reduction is not restrictive but can come from any of the following categories like polymers, solid-particle suspensions, biological additives, and surfactants. However, in each category, only some specific types of materials tend to be drag reducing. The revelation of this knowledge came from exploratory experiments and visualization studies by many researchers from chemical engineering, materials science, mechanical engineering, and polymer science.

To understand why some materials from a single category show drag reduction while others do not, researchers considered the mechanism of drag reduction. There were also attempts to explore theoretical analysis to derive expressions for the velocity profiles, friction factors, velocity distributions, and boundary layer thicknesses. Consideration was given to flow situations in various types of geometries such as the smooth/rough circular pipes, annular ducts, curved tubes, vertical flat plates, and channels.

This book provides the details of the theoretical derivations and presents expressions for fluid and convective flow of mildly elastic fluids in various flow situations involving different types of geometries. An understanding of the methodology of the analyses would be helpful to clearly see the rationale to be used for deriving expressions for the parameters relevant to the flow of mildly elastic fluids. It is hoped that with this knowledge, there would be a driving force to use the methodologies employed herein, develop new ideas, innovate new situations, and further extend the theoretical analyses to include more unexplored areas relating to the rheology of mildly elastic drag reducing fluids.

This book can be used to teach an elective course at undergraduate or graduate level to chemical engineers, materials scientists, mechanical engineers, and polymer scientists and can serve as a reference book to researchers who want to explore and expand the areas dealt in this book. Moreover, researchers who are unexposed to this fascinating area of drag reduction can find it revealing and useful.

Germantown, MD, USA Aroon Shenoy
2020

About the Book

Drag reduction is a very alluring and interesting subject. There have been tremendous research efforts towards understanding this phenomenon, and many different types of materials have been explored to get a deeper insight into the mechanism of drag reduction. Most of the knowledge in this area has evolved through experiments and visual observations, while some theoretical exploits have also been recorded. This book focuses on some of the theoretical attempts and presents expressions for fluid and convective flow of mildly elastic fluids in various flow situations involving different types of geometries. The main idea of putting forth such a compilation of theoretical derivations is to lay down a foundation for future group of researchers in the field of drag reduction to develop new ideas, innovate new situations, and build upon the methodology used for the analysis of the rheology of drag reducing fluids.

The book covers a broad range of topics from velocity profiles and friction factors to velocity distributions and boundary layer thicknesses. Fluid flow and convective flow in both external and internal flow situations have been treated, and various geometries such as the smooth/rough circular pipes, annular ducts, curved tubes, and vertical flat plates have been covered. This book can serve the needs of postgraduates from academia and industry in the fields of chemical engineering, materials science, mechanical engineering, and polymer science.

Contents

About the Author

Aroon Shenoy holds a Ph.D. in Chemical Engineering and works as a Technical Consultant. He is the Author/Coauthor of over 125 technical papers, about a dozen invited review articles/chapters in encyclopedias and handbooks, and 5 books, namely, *Heat Transfer to Non-Newtonian Fluids: Fundamentals and Analytical Expressions* (Wiley-VCH (2017)); *Convective Flow and Heat Transfer from Wavy Surfaces: Viscous Fluids, Porous Media, and Nanofluids* (with Mikhail Sheremet and Ioan Pop; CRC Press, Taylor & Francis Group, Florida (2016)); *Rheology of Filled Polymer Systems* (Kluwer Academic Publishers, Netherlands (1999)); *Selecting Thermoplastics for Engineering Applications* (with Charles P. MacDermott; Marcel Dekker Inc., New York (1997)); and *Thermoplastic Melt Rheology and Processing* (with D. R. Saini; Marcel Dekker Inc., New York (1996)).

List of Figures

List of Tables

Chapter 1
Drag Reducing Agents: A Historical Perspective

Introduction

In several practical fluid-flow situations, turbulence occurs near solid surfaces, and the main resistance to the flow of the fluid is associated with this turbulence. The energy losses due to turbulent friction are of very high magnitude, and this provides enough motivation for unabated research to find ways for its reduction. The 1930s witnessed one of the most exciting discoveries of fluid mechanics, namely, that turbulent skin friction can be reduced by the presence of certain additives in the flowing fluid. Various types of additives in the flowing fluid have been found to be effective:

(a) Macromolecules like those of polymers or surfactants
(b) Simple solids like fine grains or fibers

The discovery provided a physical means of reducing the tremendous energy losses due to turbulent skin friction in the flow of fluids through conduits or, conversely, the movement of solid bodies in fluids. The research that followed this discovery focused attention on the nature of turbulent skin friction generated at a solid surface by relative flow of a fluid past it.

Most of the researchers in this field have worked with fluids at room temperature because of the practical importance of using drag reducers in ship-building industries, for fire-fighting operations, oil-well fracturing processes, etc. where high temperatures are not involved. With the idea of using drag reducing additives in central heating systems, there have been some studies on the effectiveness of drag reducers at high temperatures. An epitome of the results of the work done on drag reduction can be obtained in many of the reviews (Berman, 1978; Graham, 2004; Hoyt, 1972a; Lumley, 1969; Nadolink & Haigh, 1995; Procaccia, L'vov, & Benzi, 2008; Virk, 1975; Wang, Yu, Zakin, & Shi, 2011; White & Mungal, 2008; Zakin, Lu, & Bewersdorff, 1998), encyclopedia chapters (Gadd, 1971a; Hoyt, 1986), monograph (Gyr & Bewersdorff, 1995), reports (Darby, 1972; Fisher & Ash, 1974; Landahl, 1972; Palyvos, 1974;

A. Shenoy, *Rheology of Drag Reducing Fluids*,
https://doi.org/10.1007/978-3-030-40045-3_1

White & Hemmings, 1976) and journal articles (Giesekus et al., 1981; Kirdyashkin, 1977; Little et al., 1975; Lumley, 1973; Patterson, Zakin, & Rodriguez, 1969; Sellin, Hoyt, Pollert, & Scrivener, 1982; Sellin, Hoyt, & Scrivener, 1982; Shi et al., 2011; Warholic, Massah, & Hanratty, 1999; Wei & Willmarth, 1992).

Review of Past Work

Historical

The earliest works that recorded a decrease in pressure drop during turbulent flow were undertaken in the 1930s (Brautlecht & Sethi, 1933; Brecht & Heller, 1935; Forrest & Grierson, 1931) and concerned the transportation of paper pulp. This was, however, not explicitly referred to as a drag reduction phenomenon. Toms was the first to recognize (Toms, 1948) the tremendous reduction in wall shear stress caused by the addition of small amounts of linear macromolecules to a flowing fluid. He noticed that under certain conditions of turbulent pipe flow, dilute solutions of polymers required less specific energy expenditure than that required for the pure solvent. This effect of specific energy reduction is occasionally termed the *Toms effect*. Other names that this phenomenon has received include the *non-Newtonian effect* and the *viscoelastic effect*. Quite often it is known as the *Texas effect* in view of the contributions of Texas investigators. It has come to be known also as the *Texas-Toms effect*, but the least objectionable and the most suited name for it is the *drag reducing effect*.

This phenomenon was attributed to a wall effect (Toms, 1948) or a region near the wall where the polymer molecules would be absent due to their bulky size. At that time Oldroyd attempted to explain the above phenomenon (Oldroyd, 1948) by suggesting that the external constraint imposed by the tube wall may introduce a locally preferred direction in a normally isotropic material. Hence, there was a possibility of the existence of an abnormally mobile laminar sublayer of a thickness comparable with molecular dimensions.

During the Second World War, an effect like the above was observed during the flow studies of gasoline thickened with an anionic surfactant, namely, aluminum soaps. The findings of this work were, however, first published much later by Mysels (1949).

After about 10 years of dormancy since the first works in drag reduction, the year 1959 witnessed the dawn of a new era in the field of drag reduction. The work at the University of Delaware (Dodge & Metzner, 1959) was published at the same time as the work at Massachusetts Institute of Technology (Shaver & Merrill, 1959). Both noticed unusually low friction factors for certain non-Newtonian solutions like those of sodium carboxymethyl cellulose in water.

At around the same time, industrial researchers (Dever, Harbour, & Siefert, 1962 and Ousterhout & Hall, 1961) made similar observations with certain gums – the most prominent being guar gum, which is a polysaccharide derived from a plant. The gums were used to suspend sand in the sand-water mixtures utilized in oil-well fracturing operations. There were also other researchers (Pruitt & Crawford, 1963; Savins, 1964) who dominated the field of drag reduction at that time. The works (Pruitt & Crawford, 1963; Savins, 1964) revived an interest among the US Navy workers to explore the potential of the drag reducing additives for military applications. Hoyt and Fabula were among the Navy researchers, who made the first significant contributions (Hoyt & Fabula, 1963, 1964). One of the most effective drag reducers, namely, poly(ethylene oxide), was identified by them (Fabula, Hoyt, & Crawford, 1963).

In the years that followed the above initial efforts, several papers and reports appeared in various journals for rheologists, chemical engineers, polymer chemists, hydrodynamicists, and oil technologists, wherein the flow characteristics of drag reducing solutions were studied for different internal and external flow situations. In the study of internal flows, almost every pipe-flow geometry has been considered. Apart from straight, smooth, rough, or corrugated pipes, studies have been carried out in curved or helically coiled tubes (Barnes & Walters, 1968, 1969; Walters, Barnes, & Dodson, 1971), in orifices (Tomita, 1970), in tube enlargements (Little & Wiegard, 1971), in annuli (Bilgen & Boulos, 1972; Corredor, Bizhani, & Kuru, 2015; Rubin & Elata, 1971), and in open channels (Bilgen, 1971; Eckelmann, 1973). A great number of theoretical and experimental studies have been carried out for many external flow situations like around rotating disks (Bilgen, 1971; Bilgen & Vasseur, 1974; Fabula et al., 1963; Giles, 1968; Gold, Amar, & Swaidan, 1973; Goldstein, 1965; Hoyt & Fabula, 1963, 1964; Kato, Watanabe, & Ueda, 1972; Peyser, 1973), rotating cylinders (Belokon & Kalashnikov, 1971; Denn & Roisman, 1969; Elata & Tirosh, 1965; James & Gupta, 1971; Jones & Marshall, 1969; Merrill, Mickley, & Ram, 1962; Rubin & Elata, 1966; Song & Tsai, 1966), flat plates (Fruman & Sulmont, 1969; Kumar & Sylvester, 1973; Levy & Davies, 1967; Wu, 1971, 1973), and spheres (Chhabra, Uhlherr, & Boger, 1980; Ruszczycky, 1965; White, 1966a, 1967b; Williams, 1965).

Definition of Drag Reduction

A general and widely accepted definition of drag reduction was provided by Savins, who defined drag reduction as the increase in pumpability of a fluid, which is achieved on the addition of small amounts of certain materials to the fluid when it is flowing under turbulent conditions (Savins, 1961, 1964).

In symbols, drag reduction for pipe flow is given by the following expression (Hoyt, 1972a):

$$\text{Drag Reduction} = \frac{\Delta P_{\mathrm{s}} - \Delta P_{\mathrm{a}}}{\Delta P_{\mathrm{s}}} \quad \text{at constant flow rate} \tag{1.1}$$

where ΔP_s is the pressure loss due to friction in unit length of the pipe for solvent alone and ΔP_a is the pressure loss due to friction in unit length of the pipe for the solution containing the additive.

Savins also introduced the term drag ratio as a means of measure of the drag reducing ability of an additive, namely, the ratio of the pressure drop of a solution containing the additive to the pressure drop of the pure solvent at the same flow rate (Savins, 1964):

$$\text{Drag Ratio} = \frac{\Delta P_{\text{solution}}}{\Delta P_{\text{solvent}}} \quad \text{at constant flow rate} \tag{1.2}$$

A solution is categorized as drag reducing if it has a drag ratio which is less than unity.

Types of Drag Reducing Additives

There are a number of types of additives which have shown themselves to be successful drag reducers. Basically, they can be broadly classified under the following four headings:

1. Polymer
2. Solid-particle suspensions
3. Biological additives
4. Surfactants

1. *Drag reduction with polymers*

Most of the studies on drag reduction have been undertaken with solutions containing small quantities, a few parts per million, of soluble high molecular weight polymers in both aqueous and organic solvents, and they have been found to be quite effective. Table 1.1, though not exhaustive, serves as a ready guide to the extensive literature available on drag reducing polymer systems.

A study of the drag reducing ability of the above polymers has led to the conclusion that any macromolecular substance with a linear structure and high molecular weight, i.e., above 500,000, is a good drag reducer. The advantage of polymers is that they are effective at very low concentrations; but, as they are very prone to mechanical degradation in dilute solutions (Aggarwal & Porter, 1980; Brostow, 1983; Chang & Darby, 1983; Mueller & Klein, 1980; Paterson & Abernathy, 1970), they have limited practical applicability.

2. *Drag reduction with solid-particle suspensions*

The drag reduction caused by the addition of suspended solid matter has been the subject of study since the beginning of the last century. The study was initiated by the fact that turbid streams of water were found to flow faster than clear ones. One of the first comprehensive experimental works was undertaken in the 1940s (Vanoni,

Table 1.1 Drag reducing polymer solutions (Shenoy, 1984)

Solute	Solvent	References
Guar gum	Water	Giles (1968), Elata and Tirosh (1965), Marris and Wang (1965), Pruitt, Simmons, Neill, and Crawford (1964), Wells (1969), and White (1966b)
	Seawater	Ram, Finkelstein, and Elata (1967)
Polyethylene oxide	Water	Gold et al. (1973), Sanders, Henderson, and White (1973), Chhabra et al. (1980), and Pruitt and Crawford (1965)
	Seawater	Chashehin, Shalavin, and Saenko (1975), Elliot and Stow (1971), Giles and Pettit (1967), Goren and Norbury (1967), Little and Patterson (1974), and Peyser and Little (1971)
	Water containing an electrolyte	Hoyt and Fabula (1964), Little and Patterson (1974), Little (1971), and Pruitt, Rosen, and Crawford (1966)
	Hydrocarbons, e.g., benzene, trichloroethylene	Ramakrishnan and Rodriguez (1973)
	Blood transfusion fluids	Hoyt and White (1966)
Polyacrylamide	Water	Pruitt and Crawford (1965), Chashehin et al. (1975), Elliot and Stow (1971), Pruitt et al. (1966), Forester, Larson, Hyden, and Wetzel (1969), Metzner and Park (1964), Nagarajan, Davies, and Venkateswarlu (1974), and Seyer and Metzner (1967a, 1967b)
	Brine	Mel'tser, El'perin, Leventhal, and Kovalenko (1972)
Sodium carboxymethyl cellulose	Water	Dodge and Metzner (1959), Shaver and Merrill (1959), Pruitt and Crawford (1965), Ernst (1966, 1967), Kobets (1969), and Ripkin and Pilch (1963)
Hydroxyethyl cellulose	Water	Meter (1964)
Polyisobutylene	Toluene, benzene, cyclohexane	Shaver and Merrill (1959) and Hershey and Zakin (1967)
	Light mineral oil	Rodriguez, Zakin, and Patterson (1967)
	Decalin	Ram and Kadim (1970)
	Crude oil, kerosene	Ram et al. (1967) and Polishchunk, Raiskii, and Temchin (1972)
Poly(isodecyl methacrylate)	Kerosene	Holtmeyer and Chatterji (1980)
Polymethyl methacrylate	Toluene	Hershey and Zakin (1967) and Rodriguez et al. (1967)
	Monochlorobenzene	Toms (1948)
Polyvinyl alcohol	Water	Chashehin et al. (1975) and Ernst (1967)
Vinyl I	Water	Savins (1964)

(continued)

Table 1.1 (continued)

Solute	Solvent	References
Sodium polystyrene sulfonate	Water	Peyser (1973)
Tri-n-butyl-stannyl fluoride	Hexane	Evans (1974)
Polystyrene	Toluene	Peyser and Little (1971), Ramakrishnan and Rodriguez (1973), Hershey and Zakin (1967), Block, Morgan, and Walker (1974), Cox, North, and Dunlop (1974), and Nadolink (1973)
	Benzene, methyl ethyl ketone	Ramakrishnan and Rodriguez (1973)
	Cyclohexane	Peyser and Little (1971)
Dowell-APE	Kerosene	Sylvester and Smith (1979)
Potassium polyphosphate	Water containing sodium pyrophosphate	Hunston, Griffith, and Little (1973)
Flax meal	Seawater	Lummus and Randall (1964)

1946). Despite this early start, it was more than 10 years before investigators began systematic study of solid suspension drag reduction.

Broadly speaking, suspensions may be considered to be of two different types, namely:

(a) Granular or nearly spherical particles
(b) Fibers

The experimental work on drag reduction with both these types of suspensions has been reviewed (Zandi, 1967). As a quick resource to some of the available literature, the following Table 1.2 is provided.

To date the effects of solids suspended in flowing liquids have not been investigated as thoroughly as those of polymer additives, but, from the industrial point of view, suspensions of solids have an edge over that of polymers for the following two reasons:

1. The solid matter can be added and removed from the flowing liquid with great ease.
2. They suffer no mechanical degradation in most cases except in a few as noted by some workers (Hoyt, 1972b; Mejean & Boulos, 1976; Vanasse, Coupal, & Boulos, 1979).

3. *Drag reduction with biological additives*

The drag reducing ability of biological additives came to light as a serendipity. When models of prospective new ships were tested in towing tanks to determine their drag and propulsion characteristics, Hoyt was the first to notice the drastic variation in the turbulent flow properties of water in those tanks (Hoyt, 1966a; Hoyt

Table 1.2 Drag reducing solid suspension solutions (Shenoy, 1984)

Suspension	Flowing fluid	References
Sand	Water	Blatch (1906) and Vanoni and Nomicos (1960)
Coal, fly ash, clay, activated charcoal	Water	Zandi (1967)
Wood and wood pulp	Water	Bugliarello and Daily (1961), Daily and Bugliarello (1961), Mih and Parker (1967), and Robertson and Mason (1957)
Fibrous wood pulp	Solution of guar gum	Bilgen and Boulos (1973)
Emery	Water	Maude and Whitmore (1958)
Thoria	Water	Eissenberg (1964) and Thomas (1962)
Nylon fibers	Water	Bobkowicz and Gauvin (1965), Kale and Metzner (1974, 1976), Kerekes and Douglas (1972), Lee, Vaselaski, and Metzner (1974), Radin, Zakin, and Patterson (1973), and Vaselaski and Metzner (1974)
	Polymer solution	Lee et al. (1974)
	Polymer solution + aerosol OT	Kale and Metzner (1974)
Rayon fibers	Water	Radin et al. (1973) and Robertson and Chang (1967)
Asbestos fibers	Water	Paterson and Abernathy (1970), Robertson and Chang (1967), Arranga (1970), Ellis (1970), Hoyt (1972b), Van Driest (1970), Van Driest (1971), and Peyser (1973)
	Polymer solution	Lee et al. (1974)
	Aerosol OT solution	Hoyt (1972b)
Yellow dye crystals	Their mother liquor	Pirih and Swanson (1972)

& Fabula, 1963). He explained that the long-chain polysaccharides produced by living organisms during growth were the cause of this drag reduction.

The drag reduction measurements (Hoyt & Soli, 1965) initiated the study of the polysaccharides of several fresh water and marine algae (Kenis, 1968a; Kenis & Hoyt, 1971). The effect of pH (Kenis, 1968b) and temperature (Kenis, 1969) was studied on the bacterial production of drag reducing polysaccharides. It was concluded that the bacteria could be cold-loving or heat-loving but they are often found to be more effective in alkaline media. They are also found to show long-term stability and are thus superior to poly(ethylene oxide), which was used initially in towing tanks (Kenis, 1971).

Seawater slime and other freshwater biological growths have also been found to be effective drag reducers. Some of the effective biological additives are given in Table 1.3.

Table 1.3 Drag reducing biological additives (Shenoy, 1984)

Biological additive	References
Porphyridium aerugineum	Hoyt and Fabula (1963)
Porphyridium cruentum	Hoyt and Soli (1965)
Xanthomonas campestris	Hoyt and Soli (1965), Kenis (1968a), and Kenis and Hoyt (1971)
Chaetoceros didymus	Kenis (1971)
Prorocentrum micans	Hoyt (1968)
Fish slimes	Nagarajan et al. (1974), Hoyt (1968, 1974), Kobets, Zar'yalova, and Komarova (1969), Merkulov and Khotinskaya (1969), Pyatetskii and Savshenko (1969), and Rosen and Cornford (1971)

4. *Drag reduction with surfactant solutions*

"Surfactant" is a very convenient contraction of the term "surface-active agent." It connotes an organic molecule or an unformulated compound having surface-active properties. Three classes of surfactants, namely, anionic, cationic, and nonionic, have been investigated for drag reduction.

(a) *Anionic surfactants as drag reducers*

(i) Aqueous Systems

Savins carried out extensive and pioneering work on anionic surfactants as drag reducers in aqueous solutions (Savins, 1967, 1968, 1969). He made use of alkali metal and ammonium soaps and obtained (Savins, 1967) a drag reduction of 30% for 0.2% sodium oleate in water at only 0.3 m/s.

The addition of an electrolyte helped to increase the drag reduction. It was observed that the drag reduction could be raised from 45% to 82% at a fixed wall shear stress (=150 dynes/cm^2) by increasing the concentration of the electrolyte, namely, KCI from 3.5% to 10%. He explained that the presence of KCI helped in the enhancement of the association of the soap molecules and that the soap micelles, which were initially spherical in the aqueous solution (Kruyt, 1949), were rearranged under the influence of the electrolyte into cylindrical shapes which in turn formed a network of interlaced rodlike elements (Pilpel, 1954). The amount of electrolyte normally ranged from 2% to 14% to produce stable association colloids for drag reduction. The soap concentrations were of the order of 0.1%, which are considerably higher than the typical polymer concentrations.

An interesting stressed controlled drag reduction effect in the soap solutions was observed (Savins, 1967), which is contrary to what is observed in the case of polymer solutions. Savins found that the percentage reduction in drag increased with an increase in (8 V/D) up to a critical shear stress $(\tau_w)_{cr}$. However, for $(\tau_w) > (\tau_w)_{cr}$, the percentage reduction in drag began to decrease, and at some higher values of (8 V/D), the soap solution became virtually indistinguishable from the soap-free electrolyte solution. According to Savins, this occurred because of a temporary disentanglement induced by turbulent vortices and eddies in fully developed turbulent flow. He noted another interesting fact that the maximum drag reduction always occurred at the critical shear stress and the critical shear stress essentially depended on the amount of electrolyte present. He also observed the amazing reversibility of the stress controlled drag reduction phenomenon. If the shear stress was reduced from above $(\tau_w)_{cr}$ to below $(\tau_w)_{cr}$, then the bonds reformed, and the drag reducing ability of the solution was restored. This is in sharp contrast to polymer molecules, which, when broken, cause the fluid to lose its drag reducing ability permanently. The highest critical shear stress up to which effectiveness was obtained by Savins was 1000 dynes/cm^2 for a concentration of 2000 ppm of his soaps. He also showed that the soaps were resistant to mechanical degradation. He described a test on a 2500 ppm soap solution which initially showed 77.1% drag reduction as compared to 78.2% after 88 continuous hours of circulation of the fluid by a centrifugal pump at a wall shear stress of about 200 dynes/cm^2.

Savins used the jet thrust technique and detected no normal stress differences in the shear rate range of 2000–125000 s^{-1}, in an aqueous solution containing 0.2% sodium oleate, 10% KCl, and 0.6% KOH. He also did not observe any critical shear stress for "onset" in the soap systems that he studied. The critical shear stress that he alludes to is the "threshold" shear stress beyond which the drag reduction activity of the soap solutions decreases steadily and virtually becomes nil at some higher shear stress.

Savins (1968) found that temperature did affect the drag reduction ability of the soaps he investigated (Savins, 1968). For a solution of 1 part by weight of sodium oleate, 3 parts by weight of sodium chloride, 0.1 part by weight of sodium hydroxide, and 95.9 parts by weight of water, the percentage reduction in pressure drop at 30 °C was more than that at 65 °C for lower flow rates and vice versa for higher flow rates. Thus, Savins concluded that it is profitable to use a lower concentration of soap for greater efficiency if the environment temperature is lower. Undoubtedly, the utility of these soaps in industry is where long-term stability of the drag reducing effect is of importance. These conventional soaps are relatively inexpensive and mechanically stable; yet, they have limited applicability as the calcium and other insoluble soaps are precipitated out by the interaction with calcium and other ions that are generally present in tap and seawater.

(ii) Non-aqueous Systems

The earliest work on anionic surfactants in nonaqueous systems dates to the Second World War when, in the course of designing and testing flame throwers, it was felt that there was a need for information on the flow characteristics of gasoline thickened with aluminum soaps which were the constituents of napalm. K. J. Mysels and his associates investigated the various aspects of these thickened solutions and found that the pressure loss per unit length of the pipe when thickened gasoline flowed under turbulent conditions was much lower than that of the pure solvent. The findings were published much later (Mysels, 1949) and the effect was attributed to a variable viscosity with shear rate of the thickened fluid (Mysels, 1949). However, it was later suggested (Agoston et al., 1954) that there was a different mode of flow of the thickened jellylike substances and the gasoline. They observed drag reduction in a 0.3 cm diameter pipe at about 15 m/s. They also showed that in more concentrated solutions, there was a pronounced viscoelastic effect with long relaxation times and that the pressure gradients persisted for several minutes after the pumping had been terminated. Apart from Mysels (1949) and Agoston et al. (1954), an account of Mysels' experiments appeared many years later (Mysels, 1971).

In 1960, it was shown (Ousterhout & Hall, 1961) that diesel oil gelled with an unnamed soap at an unspecified concentration and yielded a drag reduction of 40–50% at about 10 m/s in 5.1- and 6.3-cm-diameter pipes.

Many years after the initial demonstration of drag reduction in nonaqueous anionic surfactant solutions, a systematic study of these systems was initiated (Radin, Zakin, & Patterson, 1969). Among the first systems studied were aluminum dioleate and aluminum palmitate soaps in toluene. Unlike the aqueous anionic surfactants, drag reduction did not occur until a concentration as high as 0.75% was reached. The viscosity at this concentration was found to be two or three times that of the solvent toluene. The viscosity was found to drop slightly when reproducibility of data was being checked, thus indicating a slight mechanical degradation. Measurements were made for solutions aged for 12 days, and it was found (Radin et al., 1969) that the reduced viscosity decreased greatly at all concentrations for every initial concentration. They also found that the aged solutions were not as sensitive to mechanical degradation as fresh solutions.

Contrary to Savins' results (1967), significant values of the first normal stress differences were observed (Radin et al., 1969) in the 1% solution of aluminum dioleate in toluene for the shear rate range 2000–30000 s^{-1}. Furthermore, they did not find the critical shear stress observed by Savins. In fact, they observed no loss in drag reduction even up to a wall shear stress of 5000 dynes/cm^2. They suggested that this was due to the forces holding the micelles together in hydrocarbon solvents being stronger than in aqueous solvents.

No time or shear degradation effects were observed (McMillan, 1970) in the concentrated solutions of aluminum disoaps. McMillan suggested an equilibrium model based on which he explained his experimental results. He observed that a permanent drag reduction in aluminum distearate cannot be obtained unless a certain minimum concentration is reached. Above this concentration, he suggested that a highly structured association colloid existed which is stable with respect to time and is recovered completely after a shear breakdown. Below this critical concentration, a metastable structure exists in the solution, which may readily be transformed into its dispersed equilibrium state by aging or by high shear or by both. Lee also observed the same phenomenon, thus confirming the above findings (Lee & Zakin, 1973).

The method of preparation of the disoap solutions strongly affected their turbulent flow characteristics (McMillan, Hershey, & Baxter, 1971). Solutions prepared by diluting more concentrated batches or prepared at temperatures of 60–80 °C showed higher viscosities and were less effective drag reducers than those prepared such that they are completely mixed. Moreover, the properties of such solutions prepared by dilution were time dependent, but the differences between such nonequilibrium states were reduced with increased aging and mixing.

It was also found that aluminum dioctoate was the best drag reducer among soaps or polymers for organic solvents (McMillan et al., 1971). Aluminum dioctoate in toluene yielded equivalent amounts of drag reduction at concentrations that were an order of magnitude lower than those for aluminum distearate in toluene (McMillan et al., 1971) or aluminum dioleate in toluene (Radin et al., 1969).

A marked difference between the anionic aqueous surfactant solutions investigated by Savins and the nonaqueous ones was that the temperature dependency of drag reduction showed opposite trends. Larger percentages of drag reduction were found at 15 °C than at 25° or 35 °C.

McMillan's light scattering studies (McMillan, 1970) showed that large agglomerates or micelles caused appreciable drag reduction by affecting both the viscosity and the viscoelasticity of the solution. The effect of aging is to increase the number and size of the micelles. It was noticed (Zakin, 1972) that aging of aluminum dioctoate solution in toluene helped to increase both the apparent viscosity and the drag reducing ability by causing a more complete dispersion and a possible increase in the agglomerate size. It was observed (Rodriguez, 1971) that the stability to aging of aluminum di-2-ethylhexanoate was superior to that of other aluminum soaps investigated. A sample which had been stored as a 2% gel in toluene for over 5 years gave nearly as good a drag reduction on dilution as the initial preparation. Sheffer investigated (Sheffer, 1948) the effect of temperature on the rate of decrease of viscosity of a 0.1% aluminum dilaurate solution in benzene. He found the rate of decrease to be lowest near room temperature, but there was a rapid fall in viscosity for either lower or higher storage temperatures.

The presence of light, moisture, and traces of other substances has a significant effect on the performance of disoap-hydrocarbon systems (Baker, Bolster, & Little, 1970). It has been observed (Sheffer, 1948) that deterioration of the soap systems was accelerated by trace amounts of water and acid. On the contrary, the observations by Zakin showed (Zakin, 1972) that aging was actually retarded by the presence of water, probably because water molecules slowed down the breakup of the agglomerates or micelles. Other hydrogen bonding compounds like crotyl chloride were also found to enhance the drag reducing ability of soap solutions, by speeding up the initial soap dispersion in the solvent (Lee & Zakin, 1973). The presence of such hydrogen bonding molecules may cause either rapid dispersion, rapid aging, or stabilization of the aluminum disoap micelles depending on the kind of interactions of the hydrogen bonding molecules with the soap.

(b) *Cationic surfactants as drag reducers*

The cationic surfactant that has been investigated for drag reduction in most detail is cetyltrimethylammonium bromide (CTAB). It was shown (Nash, 1956a, 1958) that mixtures of CTAB and naphthalene derivatives in water produced viscoelastic gels. Nash made use of the swirl decay times to find the conditions under which dilute solutions of CTAB and naphthol became viscoelastic (Nash, 1956b).

Nash showed that the optimum mixture for viscoelasticity was not far from equimolar and that there was a general lowering of the concentration limits in the presence of salt. He realized that there was a pronounced effect on varying the order of addition of the components to the mixture. He found that adding naphthol to the CTAB solutions produced stronger gels than vice versa, even though the final concentrations of the reagents were identical.

It was Gadd who suggested (Gadd, 1966b) the possibility of using the CTAB-naphthol mixture to reduce turbulent friction, because the mixture showed shear-thinning characteristics. The findings of Nash and Gadd were used (White, 1967a) to carry out turbulent drag reduction measurements in circular pipes using an aqueous equimolar solution of CTAB and 1-naphthol at a total concentration of 508 wppm. The solution was prepared by dissolving CTAB in water and then adding the naphthol, which had been previously dissolved in alcohol, drop by drop. The viscosity of this resulting complex soap solution was found to be only slightly greater than that of water. White noticed no threshold stress effect of the kind that is noticed in many dilute high polymer solutions. He found that the drag reduction terminated at some upper Reynolds number depending on the pipe diameter. The breaks in the Reynolds number curves for different pipe diameters corresponded to a constant value of shear stress, $\tau_w = 48$ dynes/cm^2. He explained this as the critical shear stress at which there was a scission of the micelles. The drag reduction ability was, however, restored below the upper Reynolds number because of the reformation of the disrupted micelles. He also did not note any permanent shear degradation over a period of several days.

The drag reducing ability of CTAB-naphthol mixtures was also investigated by other researchers (Zakin, Poreh, Brosh, & Warsharsky, 1971). They carried out tests to determine which weight ratio of CTAB to naphthol gave the highest critical shear stress. They found that the maximum shear stability occurred for the 2.1 to 1 solution.

They also investigated the effect of varying the total concentration of the complex soap solution and found that the critical shear stress increased with concentration. But as the more concentrated solutions became more viscous, they were less drag reducing at lower velocities. However, at about 0.18% total soap concentration, maximum drag reduction was observed over the whole range of flow rates.

They observed that as the temperature was increased, the critical shear stress decreased gradually for a 0.215% concentrated solution, in contrast to the sharp change observed (White, 1968) in the range of 35–40 °C for a 0.05% concentration of his complex soap solution. The solutions were tested for temperatures between 26 and 48 °C (Zakin et al., 1971). It was found that the drag reducing behavior was completely lost after the 48 °C test. This loss was observed to be permanent as this solution showed no drag reduction characteristics even when cooled to 26 °C.

Some loss of drag reduction has been reported (White, 1968) for his complex soap solutions after 5 days and found that the solutions were rendered completely ineffective after 12 days. Contrary to this, a drastic drop in the critical shear stress was observed (Zakin et al., 1971) from over 500 dynes/cm^2 to less than 200 dynes/cm^2 after 3 days accompanied by a tremendous loss in the drag reduction capacity.

Seven days later there was further loss observed, but the rate of change of critical shear stress was found to be slower.

From the two studies (White, 1968; Zakin et al., 1971), it is clear that CTAB-naphthol mixtures behave like concentrated drag reducing solutions as they show no transition region but only a gradual deviation from the extension of the laminar curve in the friction factor versus Reynolds number plots. The minimum concentration for this type of drag reducing behavior was, of course, found to increase with increasing tube diameter.

The effect of adding the CTAB-naphthol mixtures to moderately concentrated sand-water suspensions was also studied (Zakin et al., 1971). They found that the pressure drops and critical velocities in the soap suspensions flows were considerably lower than in water suspensions flows, thus giving major reductions in the energy requirements per pound of solid transported. They also noticed that there was an interaction between the surface of the solid particles and the soap, resulting in a reduction of the effective soap concentration in the suspending fluid. As before, they noticed that the solutions were quite stable to mechanical degradation. One marked advantage of cationic surfactants over the anionic ones discussed earlier is that these complex soaps do not precipitate in the presence of calcium ions. But they are expensive and degrade chemically in aqueous solutions in a matter of a few days. Though mechanically stable, they are not thermally stable and hence cannot be put to a variety of uses.

(c) *Nonionic surfactants as drag reducers*

The studies on nonionic surfactants as effective drag reducing additives have been reported (Zakin & Chang, 1972, 1974). They studied the effect of temperature, electrolyte concentration and type, surfactant concentration and composition, and the effect of mechanical shear on three nonionic surfactants formed from straight-chain alcohols and ethylene oxide moieties of proper sizes. Nonionic surfactants have an upper and a lower temperature limit for solubility in water. They noted that the nonionic surfactants studied were effective drag reducers near their upper critical solubility temperature or cloud point. This cloud point is the point at which a nonionic dissolved in water becomes turbid as the temperature is raised. With the approach of the cloud point, the micelles become larger and larger till finally a surfactant-rich phase separates out. This is possibly the result of the dehydration of the hydrophobic ether linkages in the chain, thereby causing an increase in the hydrophobic nature of the chain (Kuriyama, 1962). The cloud point is almost constant over a wide range of concentrations of the surfactants; but it is affected by other additives. Electrolytes lower the cloud point; alcohols and other polar solvents may raise it; soap and other anionic surfactants may displace it.

Zakin and Chang found that the effect of increase of temperature was to change the drag reducing behavior from "dilute" to "concentrated." "Dilute" drag reducing behavior is one in which there exists a transition period even after fully developed turbulent flow is reached before the solution shows drag reduction. When no such transition points are present, the behavior is termed as "concentrated" drag reducing behavior. The terms "dilute" and "concentrated" as defined above should not be confused with the usual meanings of these words used to refer to the amount of additive present.

They found that on increasing the concentration of the electrolyte K_3PO_4 from 0.3 to 0.4 N, they could obtain "concentrated" drag reduction from what was initially no drag reduction. On investigations of various types of electrolytes, they concluded that multivalent anions were more effective than monovalent ones and that the cations generally had very little effect.

Zakin and Chang found that 1% solutions of the commercial surfactants like Alfonic 1214 were more effective than their 0.5% solutions. The critical shear stress for mechanical degradation in the case of the former was greater than 1800 dynes/cm^2 compared to 500 dynes/cm^2 in the case of the latter. The critical shear stress for mechanical degradation in the case of nonionic surfactants is dependent on the surfactant concentration, on the electrolyte type and its concentration, and on the temperature of operation. The chemical structure of the surfactant has an important effect on its micelle size and shape (Becher, 1967) and thus in turn has a pronounced effect on the drag reduction. There is need for studies of additional structures to find the optimum for drag reduction (Zakin & Chang, 1972, 1974).

Nonionic surfactants have an advantage over other drag reducing additives. They are both mechanically and chemically stable. They do not precipitate out in the presence of calcium ions and hence can be used in all impure waters, seawater, brackish water, or concentrated brine solutions. Detailed studies were carried out (Shenoy, 1976) to exploit the efficacy of nonionic surfactants to their fullest extent, and it was shown that these types of additives have excellent potential for drag reduction at high temperatures.

The Proposed Theories of Drag Reduction

Since drag reduction is characterized by large changes in the flow caused by the presence of minute quantities of additives, the objective of drag reduction is to seek studies which can predict such an equally large effect sensitive at the dilutions mechanism involved. Several theoretical explanations have been offered, but none of them have been satisfactory and exhaustive.

In 1948, Oldroyd was the first to suggest that polymer molecules, during flow, affected the region near the wall most strongly (Oldroyd, 1948). His slip-at-the-wall theory has remained a mere engineering correlation method, rather than a fundamental answer to the mechanism of drag reduction, and will remain so until an improved method for the measurement of velocity profiles very near the wall is found. Based on Oldroyd's theories, Toms proposed the idea of a shear-thinning wall layer with an extremely low viscosity which resulted in lower friction coefficients for the drag reducing solutions than for pure solvents (Toms, 1948). But, later, looking at the rheograms of drag reducing polymer solutions (Barnes & Walters, 1968; Kenis, 1968b), it was evident that they were not shear-thinning but, in fact, Newtonian by conventional viscometry method. It was the work of Walsh which completely shattered the shear thinning theory of Toms (Walsh, 1967a). He showed that solutions of polymethacrylic acid, which is essentially a shear-thickening substance, gave considerable drag reduction. Further, it was shown (Lumley, 1964) that, as viscosity is dominated by inertial forces, turbulence could not be very sensitive to shear-induced viscosity changes. Though the shear-thinning wall effect theory crumpled, it was proposed (El'perin, Smol'skii, & Leventhal, 1967) that there was another aspect of the wall effect which could possibly explain the phenomenon of drag reduction. It was suggested that there could exist an absorbed layer of polymer molecules at the pipe wall during flow and this could lower the viscosity, create a slippage, dampen turbulence pulsations, and prevent any initiation of vorticity at the wall. Experiments confirmed the presence of an apparent absorption layer (Davies & Ponter, 1966; Little, 1967). However, it was later realized that this absorption effect was a mere experimental artifact (Elata & Poreh, 1966; Little, 1969) and though polymer molecules do adhere to clean surfaces in thin films, they have no projections into the bulk of the solution which could alter the flow properties. Thus, it is evident that there is no possible relationship between polymer absorption on surfaces and the drag reducing effect.

The existence of a non-isotropic viscosity was thought of as another possible way of explaining the phenomenon of drag reduction. It was quite probable that the viscosity was low in the flow direction and high enough in all other directions to cause significant damping in the turbulent fluctuations. Theoretical work (Elata & Poreh, 1966) showed that the differences in "normal stresses" could be a mechanism for drag reduction. But it was found (Gadd, 1966b) that for concentrations of equal drag reducing ability, poly(ethylene oxide) alone showed differences in the normal stress components, while no such difference was observed in the case of guar gum solutions. However, a viscoelastic model was developed (Patterson & Zakin, 1968) for calculating reductions in turbulent bursts. The normal stress differences were evaluated by measuring the thrust from a jet of the solutions, which of course were extremely concentrated by drag reducing standards. Though the results were compatible with results of drag reduction experiments, they could not be relied upon as the solutions used during the drag reducing experiments were much more dilute than those used for the thrust experiments. Contrary to the simple Maxwell model used (Patterson & Zakin, 1968), it was suggested (Kinnier, 1965; Prather, 1966) that a Maxwell distribution of the elements would be necessary to account for the drag reducing property of poly(ethylene oxide). It was Boggs and Thompson who found that viscoelasticity gave a destabilizing effect and concluded that transition should occur earlier though experiments showed contrary results (Boggs & Thompson, 1967). They, however, suggested that in the turbulent zone, frictional drag was a function of one-third the power of the Weissenberg number, which is the ratio of the elastic forces to the viscous forces. Later, Lockett made calculations to find that viscoelasticity did give destabilization (Lockett, 1964), but he inferred that a vorticity component was introduced in the flow direction. A new theoretical model of turbulent shear flow was developed (Black, 1969) which indicated greater sublayer stability but showed an increase in the longitudinal velocity fluctuations when polymers are present during the flow. There was a suggestion that drag reduction could be due to two effects of viscoelasticity (Ruckenstein, 1973):

1. The instantaneous shear stress at the wall being smaller for viscoelastic fluid than for the corresponding Newtonian fluid
2. The renewal of the elements of liquid along the wall taking place more slowly in the case of viscoelastic fluid than in the case of a Newtonian fluid

A detailed discussion on the relationship between drag reduction and the random surface renewal in turbulent pipe flow is available (Fortuin & Klijn, 1982).

Another possible explanation for drag reduction could involve the idea of elongational flow. It is quite likely that the additions of small quantities of polymer to a solvent lead to a substantial increase in the resistance to elongational flow, thereby resulting in less turbulent bursts and thus lowering the turbulent drag. The possibility of anisotropic viscosity effects due to coil extension of polymer molecules was suggested (Shin, 1965).

Tulin was among the first to visualize (Tulin, 1966) that polymer molecules undergo an extension during shear, whereby the strain energy from the turbulent eddies is absorbed and transmitted as elastic shear waves which, of course, eventually die off due to viscosity. Tulin verified from the experimental data (Hoyt & Fabula, 1964) that the turbulent dissipation in the polymer solution is proportional to the product of concentration and square root of molecular weight. He further postulated that the increased turbulent dissipation, which could be as high as three times that of a normal fluid, was proportional to the increase in the sublayer thickness of the boundary layer and thus resulted in drag reduction. Lumley concurred with Tulin's theory that the growth of the laminar sublayer could be explained by molecular extension, but he found by calculation that the elongation itself is very slight (Lumley, 1967).

Several experiments were performed (Cottrell, Merrill, & Smith, 1969, 1970) and found that there was only a very slight macromolecular elongation during shear flow. However, Lumley postulated that it was the molecular entanglements, which extended cross-stream and resisted the formation of streamwise vortices, that caused an increase in the sublayer thickness (Lumley, 1967). But pipe drag reduction was found with poly(ethylene oxide) solutions as dilute as 0.03 ppm (Paterson & Abernathy, 1970), and, at the molecular separation distances and volume fractions involved, it is quite unlikely that molecular entanglement could play any part in drag reduction. Nevertheless, it has been shown (Ellis, Ting, & Nadolink, 1970) that polymer solutions stored for several weeks depicted lower drag reduction than freshly mixed ones, the implication being that there has been a molecular disentanglement during storage. But as polymer solutions may degrade in more ways than one, their observations must be taken with a pinch of salt. Thus, the situation not being very clear, one can disregard the "molecular entanglement" hypothesis only with some reservations. It was found (Gadd, 1966b) that poly(ethylene oxide) solutions of different molecular weights gave the same drag reduction when their concentration was proportional to the critical concentration at each molecular weight (i.e., the computed concentration for the polymer coils to touch each other). Kinnier made similar computations (Kinnier, 1965) but used the concept of "equivalent concentration" to correlate his data. He found that, to have equal drag reduction for different molecular weight polymer solutions, one should have equal volumes of polymer based upon the hydrodynamic sphere considerations.

Pfenninger postulated that the molecular stretching and the possible breaking-up of the polymer molecules interfered with and weakened the vorticity of the disturbances or amplified boundary-layer oscillations (Pfenninger, 1967).

The molecular stretching itself is induced by the mutual interactions of convoluted vortex filaments starting as nonlinear laminar oscillations close to the wall. The drag reducing additives help to delay the breakdown of these oscillations into turbulence. The net effect is a partial absorption of the kinetic energy of the vortices, thereby allowing them to grow and penetrate further away from the wall before being unstable, thus resulting in an increased sublayer thickness and a drag reduction. Peterlin pointed out that there was an extremely high energy dissipation of a vortex when it hit a macromolecule, the latter absorbing the energy by molecular stretching (Peterlin, 1970). The polymer molecules help to stabilize the surface layer by mechanical interference which suppresses the turbulence and reduces the growth of the vortex. Further, using the idea of the turbulent bursting process (Corino & Brodkey, 1969) which is initiated in the sublayer and mainly responsible for the turbulent energy production, it was shown (Gordon, 1970a, 1970b) that polymer molecules suppressed the turbulent bursts and thereby increased the resistance to stretching. Gordon suggested that, if the ejection frequency of turbulent bursts is unchanged in polymer solutions, the resistance to stretching inhibits the magnitude of the bursts and hence the flow becomes less turbulent. Further, the measurements of Latto and Shen have shown that the presence of polymers during turbulent solvent flow reduce the size range of the turbulent eddies (Latto & Shen, 1970). There was also talk of a reversal in the relative occurrence of the streamwise and spanwise vortices in the presence of polymers (Gyr, 1974). A decrease in the velocity fluctuations in the spanwise direction was also observed earlier (Rudd, 1971, 1972). Gyr's results have also shown agreement with the findings of other researchers (Brennan, 1970; Fortuna & Hanratty, 1972). Kim et al. have suggested that the presence of ionic side groups in molecules like polyacrylamide causes an extension due to charge repulsion, thus resulting in a drag reduction effectiveness proportional to the ionic content of the solution (Kim, Little, & Ting, 1973). There is concurrence (Banijamali, Merrill, Smith, & Peebles, 1974) with the conclusions of Kim, Little, and Ting. A very comprehensive and physically plausible drag reduction mechanism has been postulated (Lumley, 1969, 1970, 1973) in which macromolecular elongations initiate a sequence of changes in mean and turbulent flow structures and cause a reduction in friction.

It was Astarita who suggested that turbulence in viscoelastic liquids is perhaps not suppressed but is less dissipative than in viscous liquids (Astarita, 1965). But Gadd was among the first to suggest that drag reduction occurred not due to reduced turbulence dissipation but rather due to a decreased production of turbulence (Gadd, 1965, 1966a). Experiments showed (Johnson & Barchi, 1968) that there was a decreased production of small eddies in a developing boundary layer containing a polymer. A comprehensive theory of drag reduction was proposed (Walsh, 1967a, 1967b) that large-scale disturbances which produce Reynolds stresses some distance downstream were, previously, small disturbances at the edge of the viscous sublayer some distance upstream. The polymer molecules help to alter the energy balance of the turbulent fluctuations close to the wall, thus allowing the viscous dissipation to destroy the disturbances. This decrease in the number of disturbances moving out from the viscous sublayer eventually alters the structure of turbulence in the outer part of the boundary layer, thus resulting in lower Reynolds stresses and hence drag reduction. Walsh's theory does not predict turbulence damping in a free flow and is based upon the assumption that the phenomenon is essentially due to the existence of a wall-boundary-layer flow. This concept was well borne out by experiments (Walters & Wells, 1971) wherein polymers were exuded from a porous walled pipe section in flowing water and drag reduction was observed occurring in the downstream sections of the pipe where the polymer was in contact with the walls and pure water was flowing through the interior of the pipe. It has also been shown (Killian, 1970) that in drag reducing polymer flows, the turbulence energy is considerably reduced at high frequencies and is essentially due to the decreased production of turbulence.

One more possible mechanism by which drag reduction can occur is the development of a resistance to vortex stretching due to the presence of the additives. Larger polymer molecules, with larger relaxation times, would probably affect the large eddies with less intense stretching rates and cause their more rapid decay (Gadd, 1968). Further support of vortex stretching inhibition, based on large elongational viscosity, has been presented in a grid turbulence analysis (Kuo & Tanner, 1972). On the other hand, it has been pointed out (Gyr, 1968) that the process of vortex stretching inhibition was effective only in the case of the smallest eddies near the wall. The remarkable resistance to elongational stress of dilute polymer solutions has been shown (Metzner & Metzner, 1970), the implication being that there is resistance to vortex stretching. Considering vortex stretching inhibition (Lacey, 1974), it was postulated that there was a net macromolecular movement toward the center of the vortices, followed by a resistance to stretching due to an interaction in the long overlap regions of aligned molecules. The phenomenon of vortex inhibition as observed (Balakrishnan & Gordon, 1971; Gordon & Balakrishnan, 1972) could be explained by a resistance to vortex stretching caused by filament formation in drag reducing polymer solution.

Virk introduced the physical notion of an elastic sublayer (Virk, 1971a, 1971b), which according to him is characteristic of the drag reduction phenomenon. He suggested that the stimulation of polymer molecules by a turbulent shear flow could create an elastic sublayer at the onset of drag reduction, which could grow with increasing drag reduction and eventually occupy the entire pipe cross section at maximum drag reduction. The idea of a maximum drag reduction asymptote was proposed (Virk, 1966; Virk, Mickley, & Smith, 1970) beyond which no further viscous drag reduction could be found experimentally. This asymptote is expressed as

$$f = \frac{0.42}{\mathrm{Re}^{0.55}} \quad \text{Maximum drag reduction asymptote} \tag{1.3}$$

as compared to

$$f = \frac{16}{\mathrm{Re}} \quad \text{Poiseuille law} \tag{1.4}$$

and

$$f = \frac{0.079}{\mathrm{Re}^{0.25}} \quad \text{Blasius law} \tag{1.5}$$

Thus, it is evident that for turbulent flow drag reduction, the position of the friction factor versus Reynolds number plot is always confined to the portion between the Blasius line and Virk's maximum drag reduction asymptote.

Another aspect concerning the mechanism of drag reduction is the development of the turbulent structures and polymer deformation over time at the start of the phenomenon. An attempt to understand drag reduction over this developing time has been done using direct numerical simulations of turbulent Couette flow of a FENE-P fluid (Pereira, Mompean, Thais, & Soares, 2017). It has been shown that the initial interactions between the mean shear flow, turbulent structures, and polymer stretching are the key to understanding the step-by-step evolution of the drag reduction. After few instants from the start of the simulation, drag reduction initially takes a negative value before increasing and reaching a maximum positive value. At the minimum level of drag reduction, polymers experience their highest deformation state, and the energy required to initially stretch them is drawn mainly from the mean shear flow, thereby causing a decrease of the drag reduction leading to the minimum and negative value. Subsequently, the polymers begin to strongly interact with the partially suppressed turbulent structures, and the drag reduction begins to increase. Part of the energy stored by the molecules is released to the mean flow, increasing the drag reduction to a maximum level and decreasing the polymer extension. Finally, drag reduction reaches an asymptotic and positive value, marking the onset of the statistical steady flow state.

The transient dynamics of turbulence growth and bursting has been studied in relation to the effects of drag-reducing polymers (Zhu, Bai, Krushelnycky, & Xi, 2019). The growth of turbulence in Newtonian fluid flow begins with the domination of intense streaky velocity patterns accompanied by a sharp increase in the Reynolds shear stress and followed by quick breakdown into high-intensity small-scale fluctuations before entering the basin of statistical turbulence. Addition of drag reducing polymers leaves the initial growth of turbulence unaffected but stabilizes the primary velocity streak-vortex structure.

Throughout this section that discusses the proposed theories of drag reduction, an attempt has been made to illustrate the various approaches to a rational explanation of the drag reduction phenomenon by the different schools of thought. Though none of the above are complete in themselves, it is quite clear that each embodies an element of truth which cannot simply be disregarded. Explanations of phenomenological nature require a deep understanding of the detailed physics of the flow, and there are works that throw light on the underlying physics of drag reduction (Graham, 2014; Varshney & Steinberg, 2018).

Morphology of Micellar and Polymeric Systems

When one compares the data for CTAB-naphthol solutions (White, 1967a) on one hand with that for poly(ethylene oxide) solutions (Gadd, 1971b) on the other, it becomes obvious that the drag reduction behavioral patterns in these two cases are different. The soap solution exhibits drag reduction at low wall shear stress values, and beyond a certain critical value, the solution is found to rapidly revert to normal turbulent behavior. On the other hand, poly(ethylene oxide) solutions exhibit relatively small drag reduction at low Reynolds numbers and increase large reduction in drag at high Reynolds numbers. These two types of behavior are a consequence of the significant morphological difference between a micellar-type drag reducer and a polymeric-type drag reducer.

In the case of micellar systems, there are many different types of structures suggested. All surfactant molecules or ions, at concentrations above a minimum value characteristic of each solvent-solute system, associate to form particles called micelles. The most common and widely supported structures for these micelles are the spherical and the lamellar ones. The spherical micelle is generally conceived as a small ball-like particle of colloidal dimensions and fairly constant in size for a given species (Hartley, 1949). These spherical micelles are said to exist only in relatively dilute solutions.

In concentrated solutions, however, the lamellar micelle is favored, the existence of which for an ionic micelle is compatible with the thermodynamic data and the double-layer theory of charged colloids (Brady, 1949). A double-layer micelle is one in which the micelle formed by 50–100 molecular size particles is lined up in two flat layers.

It was found (Debye & Anacker, 1951) that in the presence of salt, C_{14}, and C_{16}, quaternary compounds had rod-shaped micelles, wherein the ions are disposed radically about a cylindrical axis with tails pointing inward and heads pointing outward. Successive planes of these ions are disposed parallel to one another and perpendicular to the length of the rod. This model fits well with the theoretical calculations on the energy of formation of the micelles and does not preclude the existence of double-layer micelles of the same ions under different environments (Halsey, 1953). It has been postulated (Booij, 1949) that in surfactant systems, spherical micelles initially rearrange into cylindrical or rodlike micelles under the influence of an electrolyte.

The current views on micelle structure are shared by all the leading investigators in the field. They all agree that several different structures are possible and, in fact, exist. It is possible for transition from one type to another to occur. Which of these structures exists in any particular system depends entirely on the system parameters such as the chemical nature of the surfactant, the solvent, the presence of other components like salts, solubilized materials, temperature, and the concentration of each component.

Since drag reducing solutions show viscoelasticity with increased concentration of the additives, a study of the same was essential to develop an insight into the morphology of the systems. It has been suggested (Pilpel, 1966b) that, in soap solutions, viscoelasticity arises from the formation of large interlinking secondary micelles. He reasoned out that if this viscoelasticity is merely due to physical entanglement of the micellar units, then it was quite immaterial whether the units are lamellar, cylindrical, or bead-like strings of spherical micelles. An idea of the morphological parameters for micellar systems, for the conditions of maximum drag reduction activity, can be estimated (Savins, 1969) using (Pilpel, 1966a) the information of Pilpel. The shape of the micelles of a complex soap causing the Toms effect has been discussed (Myska & Simeckova, 1983) who confirm the existence of fibrillous structure.

In the case of polymeric systems, it is generally agreed that the morphology of a polymer in solution resembles a loose three-dimensional network consisting of variously extended macromolecule segments (Savins, 1967). Several studies have been made to determine the important molecular parameters of polymer molecules for drag reduction to occur (Berman, 1977, 1980; Ting, 1982; Zakin & Hunston, 1980). It has been found (Liaw, Zakin, & Patterson, 1971) that the molecules should be flexible, loosely coiled, and long-chained (Hand & Williams, 1969, 1971; Peyser & Little, 1971) with relatively small side groups (Parker & Joyce, 1974; Ting & Kim, 1973). It has also been found that of drag reduction effectiveness of a homologous series of polymers increases with increasing chain (Paterson & Abernathy, 1970; Whitsitt, Harrington, & Crawford, 1968) and that as the highest molecular weight components disappear, the polymer solution becomes less effective.

The morphological difference between micellar and polymeric systems explains the drag reducing behavior of the two systems. These two kinds of behavior can be reconciled if one can realize that:

(a) The flexible polymer molecule needs to be elongated at a large velocity gradient before its full drag reducing ability is developed
(b) The surfactant particles are oriented much more easily at lower velocity gradients but are broken down at the high shear stresses associated with large gradients

Further, in terms of equivalent molecular weight, micelles are known to have larger values (Hershey, Kuo, & McMillan, 1975) than polymers, and hence they would naturally shift the onset of drag reduction to a lower shear stress value.

The Potential of Drag Reduction

Even after decades of research since Toms discovered the phenomenon of drag reduction, this effect has not been put into wide practical use, though a few cases of pragmatic value have been recorded over the years. This section outlines the areas where steps have been taken to put the drag reduction phenomenon into effect, thereby giving an idea of its potential.

It is a common practice to use guar gum in the hydraulic fracturing of oil wells to obtain a reduction in the turbulent friction. It has been found (Pruitt et al., 1964) that by using an optimum concentration of guar gum, determined by a nomogram as a function of well depth, pumping rate, and total volume of the fluid to be pumped, the horsepower required during the operation could be reduced by a factor of over five as compared with the use of water alone. Extensive data has also been provided (Melton & Malone, 1974) on the estimated friction resistance for oil-well pipelines with different additive concentrations.

One of the most promising applications of friction reduction has been in fire-fighting operations (Fabula, 1971; Rubin, 1972; Thorne, 1974; Thorne, Theobald, & Mahendran, 1975). It has been ascertained that addition of small quantities of drag reducers causes a considerable reduction in pressure drop through the hose, increases the flow volume, and furthers the throw of the water stream (Green, 1971; Scott, 1969). An emergency increase in the system capacity can also be achieved in times of high demand like a major fire by addition of drag reducers to the entire city water supply system (Jackson & Mayer, 1970).

Irrigation systems offer an equally promising area of application, as it has been found that an irrigation pump can cover twice as much area with the presence of drag reducing additives in the supply system (Elias & Vocel, 1978; Fajzullaev, 1974). The drag reduction studies in open channels (Sellin & Barnard, 1970, 1971) can be efficiently applied to hydraulic problems like coping up with peak discharges during flooding of waterways. The practical use of polymeric drag reducers has been seen in Dallas, USA, and Bristol, England (Sellin, 1978; Sellin & Ollis, 1980), where they are injected into storm sewers at times of peak loads during heavy storms to handle the increased runoff.

It has been shown (Ram et al., 1967) that addition of high molecular weight polyisobutylene greatly reduced the turbulent friction in the transportation of crude oil and kerosene. The successful tests of polymer injection into the Alaska pipeline to increase the throughput of Alaska crude (Burger, Chorn, & Perkins, 1980; Burger, Munk, & Wahl, 1980) demonstrate the pragmatic use of the drag reduction phenomenon. It has been shown (Poreh, Zakin, Brosh, & Warsharsky, 1970) that the presence of drag reducers considerably reduces the power levels in the transportation of suspensions of sand and slag particles. Field test in Czechoslovakia using drag reducers for hydro-transport of fly ash from power plants burning high ash coals (Pollert, 1977) is also another example of the practical use of drag reducing agents. The only factor limiting the widespread use of polymers as drag reducers in increasing the capacity of pipelines is the problem of degradation (Treiber &

Sieracki, 1970). But, in the case of oil pipelines which happen to be in cold areas, the drag reducers not only decrease the pumping costs but also retard the fall in oil temperature by lowering the liquid-pipe wall heat transfer coefficient.

In the ship-building industry, drag reducers find good use in the model testing of prospective new ships (Emerson, 1965; Hoyt, 1966a; Kowalski, 1966; Tothill, 1967). Their presence in towing tanks has a twofold advantage – first, by being able to simultaneously scale the Froude and Reynolds numbers, the operating conditions are improved, and second, by lowering the drag on the model, they help to bring the wave-making resistance coefficient of the model to the same as that of the full-scale ship.

The concept of drag reduction in external flow situations can be effectively used for reducing the drag on full-scale ships and other submerged vessels (Dove, 1966; Gollan, Tulin, & Rudy, 1970; Lang, 1969; Oltmann, 1969; Vogel & Patterson, 1964; Wells, 1969). Since large quantities of additives are required (Hoyt & Fabula, 1964; Kowalski, 1968a), the economics depend on finding very cheap drag reducers. Nevertheless, pulse ejection of the drag reducers has been suggested (Kowalski, 1968b) to reduce the additive requirements. It has been pointed out (Wu, 1969) that ejecting dilute drag reducing solutions in the sublayer around a traveling ship could produce extremely economical results. Full-scale trials have been reported (Canham, Catchpole, & Long, 1971) on the 140-feet-long ship, the HMS Highburton, showing very encouraging results. The tremendous research in this area will certainly lead to the eventual commercial use of drag reduction for ships.

There has been an interest in using drag reducing additives as possible circulation aids for diseases such as arteriosclerosis and coronary thrombosis during which the blood flow is turbulent (Greene, Nokes, & Thomas, 1970, 1971; Greene, Thomas, Mostordi, & Nokes, 1974). Hoyt and White (1966) It has been suggested (Hoyt & White, 1966) that poly(ethylene oxide) with plasma and dextran can be used to reduce the pressure drop in cardiovascular systems, provided they have no adverse physiological effects. Hoyt has patented the use of additives with blood transfusion fluids (Hoyt, 1971). It has been found that okra gum, apart from being a good drag reducer, is a successful blood plasma replacement like dextran and has a considerable resistance to shear degradation (Castro & Neuwirth, 1971). Several other in vitro experiments have been undertaken (Greene, 1972), and these have been followed by several toxicity studies (Nokes, Greene, & Thomas, 1971) which obviously are the first steps to in vivo experimentations.

Certain sophisticated experiments have been performed (Brady, 1949; Killen, 1972) with typical drag reducers, and it was concluded that their presence causes a marked reduction in flow noise radiated from the boundary layer in the high-frequency range. A surface pressure fluctuation decrease was also found to occur due to the presence of these additives (Barker, 1973; Killen & Almo, 1969). The drag reducers cause a suppression of the turbulent bursts in the boundary layer, and hence they can be used effectively to decrease the noise levels.

Applied research has shown that the presence of drag reducing additives improves the efficiency and head characteristics of turbomachines (Latto & Czaban,

1974). For example, in the case of large single-suction, centrifugal pumps, the backing plate drag can be reduced by as much as 50% using drag reducers. It has been shown (Hampson & Naylor, 1975) that high molecular weight polymers could be used for reducing friction in hydrodynamic journal bearings.

Hydrofoils are widely used (Fruman, Sundaram, & Daugard, 1974) among others like keels, rudders, propeller blades, stabilizing fins, or lifting surfaces for hydrofoil boats, and one of the problems is to increase the lift-to-drag-ratio. However, it has been found (Lehmann & Suessmann, 1972) that the lift problems in hydrofoils can be solved using drag reducers, but special hydrofoil designs and special ejection processes are essential, lest there should be large reductions of lift probably due to a contracted wake induced by viscoelasticity (Kawada & Tagori, 1973; Sarpkaya, 1973; Wolff & Cahn, 1971). Propellers at low Reynolds numbers have also been tested (Sanders et al., 1973) and comparable results (Sarpkaya, 1973) reported.

It has been found that typical drag reducing additives inhibit the tendency of water to form a vortex in free draining flows (Balakrishnan & Gordon, 1971; Gordon & Balakrishnan, 1972). This phenomenon of vortex inhibition has been shown to correlate very well with the drag reducing ability of the additives and hence could be used to predict the effectiveness of an additive as a drag reducer (Balakrishnan & Gordon, 1971). It has also been shown (Gordon & Balakrishnan, 1972) that this effect could be used in molecular weight distribution studies.

The phenomenon of friction reduction could be useful in studying molecular properties like molecular configurations (Hoyt, 1966b) and determination of molecular weight (Hoyt, 1966c). It has been suggested (Kenis, 1968a) that it could also be used for following polymerization – depolymerization reactions in biological systems. The findings (Beattie, 1974; Greskovich & Shries, 1971), that polymers provide a reduction in friction even in two-phase systems like a gas-liquid one, could be put to very good use.

Drag reducers could be put into excellent use in central heating systems, where the main resistance to heat flow is not on the liquid side and hence the presence of the additives may not have too much of an inhibiting effect upon the heat transfer. The use of polymer drag reduction additives in domestic central heating systems has been studied (Fitzgerald, 1967), but it was concluded that they are not suitable mainly because of the problem of polymer degradation. The problem of degradation is quite acute in central heating systems as the additives are liable to both mechanical and thermal degradation.

The possibility of using micellar systems as drag reducers for this application has been considered (Ahrnborn & Hagstrand, 1977; Shenoy, 1976). However, drag reduced flow, unfortunately, is accompanied by lower convective heat transfer coefficients, which of course is undesirable in district heating and cooling systems, as well as for other applications like heated tube bundles for undersea petroleum production, and other recirculating heat transport systems.

A work-around for this has been suggested and demonstrated (Maxson, Watson, Karandikar, & Zakin, 2017) using rotating agitators that are installed inside the inner tube of concentric tube heat exchangers resulting in an enhancement of the heat transfer during turbulent drag reducing flows of surfactant solutions. It was found that the agitation could increase the inner heat transfer coefficient to exceed that of pure water.

Another work-around has also been suggested (Kotenko, Oskarsson, Bojesen, & Nielsen, 2019) by shifting to heat supply based on renewable energy sources and small-scale biogas CHP plants which no longer require shell and tube heat exchangers. This opens opportunities for new surfactants and its application in district heating. The research article focuses on empirical laboratory scale study of the two drag reducing products developed by AkzoNobel (Sweden) specifically for aqueous solutions for different temperature ranges.

With elapse of time, the potential of drag reduction is bound to be fully exploited as is evident from the developments so far.

References

Aggarwal, S. H., & Porter, R. S. (1980). Shear degradation of poly(vinyl acetate) in toluene solutions by high-speed stirring. *Journal of Applied Polymer Science, 25*(2), 173–185.

Agoston, G. A., Harte, H. W., Hottel, H. C., Klemm, W. A., Mysels, K. J., Pomeroy, H. H., et al. (1954). Flow of gasoline thickened by napalm. *Industrial and Engineering Chemistry, 46*(5), 1017–1019.

Ahrnborn, L., & Hagstrand, U. (1977). *Toms effect in district heating tube systems*. Studsvik Report SVF-50 in Swedish.

Arranga, A. B. (1970). Friction reduction Charcteristics of fibrous and colloidal substances. *Nature, 225*(5231), 447–449.

Astarita, G. (1965). Possible interpretation of the mechanism of drag reduction in viscoelastic liquids. *Industrial & Engineering Chemistry Fundamentals, 4*(3), 354–356.

Baker, H. R., Bolster, N. N., & Little, R. C. (1970). Association colloids in nonaqueous fluids – viscosity and drag reduction characteristics. *Industrial & Engineering Chemistry Product Research and Development, 9*(4), 541–548.

Balakrishnan, C., & Gordon, R. J. (1971). A new viscoelastic phenomenon with significance in turbulent drag reduction. *Nature Physical Sciences, 231*, 177–180.

Banijamali, S. H., Merrill, E. W., Smith, K. A., & Peebles Jr., L. H. (1974). Turbulent drag reduction by Polyacrylic acid. *AICHE Journal, 20*(4), 824–827.

Barker, S. J. (1973). Radiated noise from turbulent boundary layers in dilute polymer solutions. *Physics Fluids, 16*(9), 1387–1394.

Barnes, H. A., & Walters, K. (1968). Dynamic similarity and drag reduction in flow of elastic liquids through curved pipes. *Nature, 219*(5149), 57–59.

Barnes, H. A., & Walters, K. (1969). On the flow of viscous and elastico-viscous liquids through straight and curved pipes. *Proceedings of the Royal Society of London, 314*(1516), 85.

Beattie, D. R. (1974). Drag reduction phenomena in gas-liquid systems. In *Proc. intern. conf. on drag reduction*. Cambridge, UK.

Becher, P. (1967). In M. J. Schick (Ed.), *Nonionic surfactants, micelle formation in aqueous and nonaqueous solutions* (p. 496). New York: Marcel Dekker.

Belokon, V. S., & Kalashnikov, V. N. (1971). Polymer additives and turbulent friction near rough surfaces. *Nature Physical Sciences, 229*(2), 55–56.

Berman, N. S. (1977). Drag reduction of the highest molecular weight fractions of polyethylene oxide. *Physics of Fluids, 20*(5), 715–718.

Berman, N. S. (1978). Drag reduction by polymers. *The Annual Review of Fluid Mechanics, 10*, 47–64.
Berman, N. S. (1980). Evidence of molecular interactions in drag reduction in turbulent pipe flows. *Polymer Engineering & Science, 20*(7), 451–455.
Bilgen, E. (1971). *Fifth intern. conf. on fluid scaling. H4*. Coventry, UK: University of Warwick.
Bilgen, E., & Boulos, R. (1972). Turbulent flow of drag reducing fluids between concentric rotating cylinders. *Transactions of the American Society of Mechanical Engineers, 1*, 25–30.
Bilgen, E., & Boulos, R. (1973). Friction reduction by chemical additives in the turbulent flow of fibre suspensions. *The Canadian Journal of Chemical Engineering, 51*(4), 405–411.
Bilgen, E., & Vasseur, P. (1974). On the friction reducing non-Newtonian flow around an enclosed disk. *Transactions ASME Journal of Applied Mechanics, 41*(1), 45–50.
Black, T. J. (1969). Viscous drag reduction examined in the light of a new model of wall turbulence. In C. S. Wells (Ed.), *Viscous drag reduction* (pp. 383–407). New York: Plenum Press.
Blatch, N. S. (1906). Water filtration at Washington D. C. *Transactions ASCE, 57*, 400–408.
Block, H., Morgan, A. M., & Walker, S. M. (1974). The reduction of turbulent drag and the degradation of polystyrene in toluene. In *Intern. conf. on drag reduction*. Cambridge, UK.
Bobkowicz, A. J., & Gauvin, W. H. (1965). Turbulent flow characteristics of model fiber suspension. *The Canadian Journal of Chemical Engineering, 43*(2), 87–91.
Boggs, F. W., & Thompson, J. (1967). *Flow properties of dilute solutions of polymer, part I – Mechanism of drag reduction, and part III – Effect of solute on turbulent field.* US Rubber Co. Research Center Report on Contract Nos. Nonr-3120(00) and N00014-66-C0332.
Booij, H. L. (1949). In H. R. Kruyt. Colloid science II, association colloids, Ch. 14, Elsevier Publishing Company, Amsterdam, pp. 681–722.
Brady, A. P. (1949). The diffuse ionic layer in relation to lamellar micelles in aqueous solutions of colloidal electrolytes. *The Journal of Chemical Physics, 53*(6), 947–955.
Brautlecht, C. A., & Sethi, J. R. (1933). Flow of paper pulps in pipelines. *Industrial and Engineering Chemistry, 25*(3), 283–288.
Brecht, W., & Heller, H. (1935). Der rohrreibungsverlust von stoffaufschwemmungen. *Wochenblatt für Papierfabrikation, 16*, 264, 342, 380, 439, 474, 529, 587, 641, 714, 747.
Brennan, C. (1970). Some cavitation experiments with dilute polymer solutions. *Journal of Fluid Mechanics, 44*(1), 51–63.
Brostow, W. (1983). Drag reduction and mechanical degradation in polymer solutions in flow. *Polymer, 24*(5), 631–638.
Bugliarello, G., & Daily, J. W. (1961). Rheological models and laminar shear flow of fiber suspensions. *TAPPI, 44*, 881–893.
Burger, E. D., Chorn, L. G., & Perkins, T. K. (1980). Studies of drag reduction conducted over a broad range of pipeline conditions when flowing Prudhoe Bay crude-oil. *Journal of Rheology, 24*(5), 603–626.
Burger, E. D., Munk, W. R., & Wahl, H. A. (1980). *Flow increase in the trans Alaska pipeline using a polymeric drag reducing additive.* Paper SPE 9419, 55th Annual Fall Conference of Society of Petroleum Engineers of AIME. Dallas, TX, September 21–24.
Canham, H. J., Catchpole, J. P., & Long, R. F. (1971). Boundary layer additives to reduce ship resistance. *The Naval Architect Journal, 2*, 187–213.
Castro, W. E., & Neuwirth, J. G. (1971). Reducing fluid friction with okra. *Chemische Technik, 1*, 697.
Chang, H. F., & Darby, R. (1983). Effect of shear degradation on the rheological properties of dilute drag-reducing polymer solutions. *Journal of Rheology, 27*(1), 77–88.
Chashehin, I. P., Shalavin, N. T., & Saenko, V. A. (1975). Effect of polymeric additives on drag reduction. *International Journal of Chemical Engineering, 15*, 88.
Chhabra, R. P., Uhlherr, P. H., & Boger, D. V. (1980). The influence of fluid elasticity on the drag coefficient for creeping flow around a sphere. *Journal of Non-Newtonian Fluid Mechanics, 6* (3–4), 187–199.

Corino, E. R., & Brodkey, R. S. (1969). A visual investigation of the wall region in turbulent flow. *Journal of Fluid Mechanics, 37*(1), 1–30.

Corredor, F. E., Bizhani, M., & Kuru, E. (2015). Experimental investigation of drag reducing fluid flow in annular geometry using particle image velocimetry technique. *Transactions ASME Journal Fluid Engineering (online), 137*(8), 08113 (16).

Cottrell, F. R., Merrill, E. W., & Smith, K. A. (1969). Conformation of polyisobutylene in dilute solution subjected to a hydrodynamic shear field. *Journal of Polymer Science Series A-2, 7*(8), 1415–1434.

Cottrell, F. R., Merrill, E. W., & Smith, K. A. (1970). Intrinsic viscosity and axial extension ratio of random coiling macromolecules in a hydrodynamic shear field. *Journal of Polymer Science Series A-2, 8*(2), 289–294.

Cox, L. R., North, A. M., & Dunlop, E. H. (1974). *Intern. conf. on drag reduction*. Cambridge, UK: BHRA.

Daily, J. W., & Bugliarello, G. (1961). Basic data for dilute fiber suspensionsin uniform flow with shear. *TAPPI, 44*, 497–512.

Darby, R. (1972). *A review and evaluation of drag reduction theories*. Naval Research Lab., NRL Memo Report 2446.

Davies, G. A., & Ponter, A. B. (1966). Turbulent flow properties of dilute polymer solutions. *Nature, 212*(5057), 66.

Debye, P., & Anacker, E. W. (1951). Micelle shape from disymmetry measurements. *The Journal of Physical and Colloid Chemistry, 55*(5), 644–655.

Denn, M. M., & Roisman, J. J. (1969). Rotational stability and measurement of normal stress functions in dilute polymer solutions. *AICHE Journal, 15*(3), 454–459.

Dever, C. D., Harbour, R. J., & Siefert, W. F. (1962). Method of decreasing friction loss in flowing liquids. US Patent 3,023,760.

Dodge, D. W., & Metzner, A. B. (1959). Turbulent flow of non-newtonian systems. *AICHE Journal, 5*(2), 189–204.

Dove, H. L. (1966). The effect on resistance of polymer additives injected into a boundary layer of a frigate model. In *Proc. 11th intern. towing tank conf.* Tokyo, Japan.

Eckelmann, H. (1973). *Bericht Nr* (p. 101). Gottingen, Germany: Max-Planck-Institut fur Stromungsforchung.

Eissenberg, D. M. (1964). Measurement and correlation of turbulent friction factors of thoria suspensions at elevated temperatures. *AICHE Journal, 10*(3), 403–407.

Elata, C., & Poreh, M. (1966). Momentum transfer in turbulent shear flow of an elastico-viscous fluid. *Rheologica Acta, 5*(2), 148–151.

Elata, C., & Tirosh, J. (1965). Frictional drag reduction. *Israel Journal of Technology, 3*(1), 1–6.

Elias, V., & Vocel, J. (1978). Vodohospodarsky eas. *SAV, 26*, 610.

Elliot, J. H., & Stow, F. S. (1971). Solutions of drag-reducing polymers – diameter effect and rheological properties. *Journal of Applied Polymer Science, 15*(11), 2743.

Ellis, A. T., Ting, R. Y., & Nadolink, R. H. (1970). *Some effects of storage and shear history on the friction reducing properties of dilute polymer solutions.* AIAA Paper 70–532.

Ellis, H. D. (1970). Effects of shear treatment of drag-reducing polymer solutions and fibre suspensions. *Nature, 226*(5243), 352–353.

El'perin, I. T., Smol'skii, B. M., & Leventhal, L. I. (1967). Decreasing the hydrodynamic resistance of pipelines. *International Journal of Chemical Engineering, 7*, 276–295.

Emerson, A. (1965). Model experiments using dilute polymer solutions instead of water. *Transactions North East Coast Institution of Engineers & Shipbuilders, 81*, 201.

Ernst, W. D. (1966). Investigation of turbulent shear flow of dilute aqueous CMC solutions. *AICHE Journal, 12*(3), 581–586.

Ernst, W. D. (1967). Turbulent flow of elasticoviscous non-newtonian fluid. *AIAA Journal, 5*(5), 906–909.

Evans, A. P. (1974). A new drag-reducing polymer with improved shear stability for nonaqueous systems. *Journal of Applied Polymer Science, 18*(7), 1919–1925.

Fabula, A. G. (1971). Fire-fighting benefits of polymeric friction reduction. *Transactions ASME Journal of Basic Engineering, 93*(3), 453–455.

Fabula, A. G., Hoyt, J. W., & Crawford, H. R. (1963). Turbulent flow characterisitcs of dilute aqueous solutions of high polymers. *The Bulletin of the American Physical Society, 8*, 15.

Fajzullaev, D. P. (1974). Uvelicenije raschoda vody v trubo-provdach s pomose ju maloj dobavky preparata E-1, *Doklady AV UzSSR*, 7.

Fisher, M. C., & Ash, R. L. (1974). A general review of the concepts for reducing skin friction, including recommendations for future studies. *NASA TMX, 2894*, L-9119.

Fitzgerald, D. (1967). *Brit. soc. of rheol. symp. on non-newtonian flow through pipes and passages.* Shrivenham, UK.

Forester, R. H., Larson, R. F., Hyden, J. W., & Wetzel, J. M. (1969). Effects of polymer addition on friction in a 10-inch diameter pipe. *Journal of Hydronautics, 3*, 59.

Forrest, F., & Grierson, G. A. (1931). Friction losses in cast iron pipe carrying papaer stock. *Paper Trade Journal, 92*, 39–41.

Fortuin, J. M., & Klijn, P. J. (1982). Drag reduction and random surface renewal in turbulent pipe flow. *Chemical Engineering Science, 37*(4), 611–623.

Fortuna, G., & Hanratty, T. J. (1972). The influence of drag reducing polymers in turbulence in the viscous sublayer. *Journal of Fluid Mechanics, 53*(3), 575–586.

Fruman, D. H., Sundaram, T. R., & Daugard, S. J. (1974). Effect of drag-reducing polymer injection on the lift and drag of a two-dimensional hydrofoil. In *Proc. intern. conf. on drag reduction.* Cambridge, UK: Paper E2.

Fruman, D., & Sulmont, P. (1969). Reduction de la resistance de prottement d'une plaque plane dans les solutions de polymeres. *Comptes Rendus. Académie des Sciences, 268*, 1493.

Gadd, G. E. (1965). Turbulence damping and drag reduction produced by certain additives in water. *Nature, 206*(4983), 463–467.

Gadd, G. E. (1966a). Reduction of turbulent friction in liquids by dissolved additives. *Nature, 212* (5065), 874–877.

Gadd, G. E. (1966b). Differences in normal stress in aqueous solutions of turbulent drag reducing additives. *Nature, 212*(5069), 1348–1350.

Gadd, G. E. (1968). Effects of drag-reducing additives on vortex stretching. *Nature, 217*(5133), 1040–1042.

Gadd, G. E. (1971a). Friction reduction. In *Encyclopedia of polymer science and technology* (Vol. vol. 15). New York: Wiley.

Gadd, G. E. (1971b). Reduction of turbulent drag in liquids. *Nature, 230*(10), 29–31.

Giesekus, H., Bewersdorff, H. W., Dembek, G., Kwade, M., Martischius, F. D., & Scharf, R. (1981). Fortschritte der. *Verfahrenstechnik, 19*, 3.

Giles, W. B. (1968). Similarity laws of friction-reduced flows. *Journal of Hydronautics, 2*(1), 34–40.

Giles, W. B., & Pettit, W. T. (1967). Stability of dilute viscoelastic flow. *Nature, 216*(5114), 470–472.

Gold, P. T., Amar, P. K., & Swaidan, B. E. (1973). Friction reduction degradation in dilute poly (ethylene oxide) solutions. *Journal of Applied Polymer Science, 17*(2), 333–350.

Goldstein, S. (1965). On the resistance to the rotation of a disc immersed in a fluid. *Proceedings of the Cambridge Philosophical Society, 31*, 232.

Gollan, A., Tulin, M. P., & Rudy, S. L. (1970). *Development and model tests of a surface ship additive system* (Report 909-1). Hydronautics Inc. Tech.

Gordon, R. J. (1970a). On the explanation and correlation of turbulent drag reduction in dilute macromolecular solutions. *Journal of Applied Polymer Science, 14*(8), 2097–2105.

Gordon, R. J. (1970b). Mechanism for turbulent drag reduction in dilute polymer solutions. *Nature, 227*(5258), 599–600.

Gordon, R. J., & Balakrishnan, C. (1972). Vortex inhibitions: A new viscoelastic effect with importance in drag reduction and polymer characterisation. *Journal of Applied Polymer Science, 16*(7), 1629–1639.

Goren, Y., & Norbury, J. F. (1967). Turbulent flow of dilute aqueous polymer solutions. *Transactions ASME Journal of Basic Engineering, 89*(4), 814–822.

Graham, M. D. (2004). Drag reduction in turbulent flow of polymer solutions. *Rheology Reviews, 2*, 143–170.

Graham, M. D. (2014). Drag reduction and the dynamics of turbulence in simple and complex fluid. *Physics of Fluids, 26*, 101301.

Green, J. H. (1971). Effect of polymer additives on nozzle stream coherence: A preliminary study, Navel Undersea R and D Center TN 504.

Greene, H. L. (1972). *Proc. 25th ACEMB*. Bal Harbour, FL.

Greene, H. L., Nokes, R. F., & Thomas, L. C. (1970). Drag reduction in pulsed flow of blood. *Research in Medical & Engineering, 9*, 19.

Greene, H. L., Nokes, R. F., & Thomas, L. C. (1971). Drag reduction phenomena in pulsed blood flow. In *ASME symposium on flow*. Pittsburgh, PA: Paper 4-4-64.

Greene, H. L., Thomas, L. C., Mostordi, E. A., & Nokes, R. F. (1974). Potential biomedical applications of drag reducing agents. In *Proc. intern. conf. on drag reduction*. Cambridge, UK.

Greskovich, E. J., & Shries, A. L. (1971). Drag reduction in two-phase flows. *Industrial & Engineering Chemistry Fundamentals, 10*(4), 646–648.

Gyr, A. (1968). Analogy between vortex stretching by drag-reducing additives and vortex stretching by fine suspensions. *Nature, 219*(5157), 928–929.

Gyr, A. (1974). *ETH Zurich: Inst. fur hydromechanik und wasserwirtschaft* (Report R7-74).

Gyr, A., & Bewersdorff, H.-W. (1995). *Drag reduction of turbulent flows by additives*. Dordrecht, Netherlands: Kluwer Academic Publisher.

Halsey, G. D. (1953). On the structure of micelles. *The Journal of Physical Chemistry, 57*(1), 87–89.

Hampson, L. G., & Naylor, H. (1975). Friction reduction in journal bearings by high molecular weight polymers. In *Proc. of the 2nd leads-lyon symposium on tribology* (pp. 70–72). London: Mechanical Engineering Publications Ltd.

Hand, J. H., & Williams, M. C. (1969). Effect of secondary polymer structure on the drag-reducing phenomenon. *Journal of Applied Polymer Science, 13*(11), 2499–2503.

Hand, J. H., & Williams, M. C. (1971). The role of polymer conformation in drag reduction. *AICHE Chemical Engineering Progress Symposium Series No. 111, 67*, 6.

Hartley, G. S. (1949). Organised structure in soap solutions. *Nature, 163*(4150), 767–768.

Hershey, H. C., & Zakin, J. L. (1967). Existence of two types of drag reduction in pipe flow of dilute polymer solutions. *Journal of Industrial and Engineering Chemistry, 6*(3), 381–387.

Hershey, H. C., Kuo, J. T., & McMillan, M. L. (1975). Drag reduction of straight and branched chain aluminum disoaps. *Industrial and Engineering Chemistry Product Research and Development, 14*(3), 192–199.

Holtmeyer, M. D., & Chatterji, J. (1980). Study of oil soluble polymers as drag reducers. *Polymer Engineering & Science, 20*(7), 473–477.

Hoyt, J. W. (1966a). The use of Porphyridium Aerugineum as a sealing aid in towing tanks. In *Proc. 11th inter. towing tank conf.* Tokyo.

Hoyt, J. W. (1966b). Turbulent flow properties of deoxyribonucleic acid solutions. *Nature, 211* (5045), 170–171.

Hoyt, J. W. (1966c). Friction reduction as an estimator of molecular weight. *Journal of Polymer Science Part B, 4*(10), 713–716.

Hoyt, J. W. (1968). Turbulent flow properties of polysaccharide solutions. In *Solution properties of natural polymers, 207, special publication N. 23*. London: The Chemical Society.

Hoyt, J. W. (1971). *Blood transfusion fluids having reduced turbulent friction properties.* US Patent No. 3,590,124.

Hoyt, J. W. (1972a). The effect of additives on fluid friction. *Transactions ASME Journal of Basic Engineering, 94*(2), 258–285.

Hoyt, J. W. (1972b). *Turbulent flow of drag-reducing suspensions* (Report NUC TP 299). Naval Undersea Center.

Hoyt, J. W. (1974). Hydrodynamic drag reduction due to fish slimes. In *Symposium on swimming and flying in nature. California institute of technology*. New York: Plenum Press.
Hoyt, J. W. (1986). Drag reduction. In *Encyclopedia of polymer science and engineering* (Vol. vol. 5, p. 129). New York: Wiley.
Hoyt, J. W., & Fabula, A. G. (1963). Frictional resistance in towing tanks. In *Proc. 10th intern. towing tank conf.* Teddington, UK.
Hoyt, J. W., & Fabula, A. G. (1964). The effect of additives on fluid friction. In *Proc. 5th symp. on naval hydrodynamics* (Vol. 112, p. 947). Bergen, Norway: Office of Naval Research.
Hoyt, J. W., & Soli, G. (1965). Algal cultures: Ability to reduce turbulent friction in flow. *Science, 149*(3691), 1509–1511.
Hoyt, J. W., & White, W. D. (1966). High polymer additive on turbulent flow of dextran, saline solution and plasma. In *Proc. 19th annual conf. in engg. on medicine and biology* (p. 49).
Hunston, D. L., Griffith, J. R., & Little, R. C. (1973). Drag reducing properties of polyphosphates. *Nature Physical Sciences, 245*(148), 140–141.
Jackson, H. C., & Mayer, P. G. (1970). *Georgia institute of technology: Unsteady flow of dilute aqueous polymer solutions in pipe networks – A method to improve water distribution.* Water Resources Center Report WRC 0170.
James, D. F., & Gupta, O. P. (1971). Drag on circular cylinders in dilute polymer solutions, AIChE. *Chemical Engineering Progress Symposium Series No. 111, 67*, 62.
Johnson, B., & Barchi, R. H. (1968). Effect of drag-reducing additives on boundary layer turbulence. *Journal of Hydronautics, 2*(3), 168–175.
Jones, W. M., & Marshall, D. E. (1969). Relaxation effects in couette flow between rotating cylinders. *Journal of Physics D, 2*(6), 809–814.
Kale, D. D., & Metzner, A. B. (1974). Turbulent drag reduction in fiber-polymer systems: Specificity considerations. *AICHE Journal, 20*(6), 1218–1219.
Kale, D. D., & Metzner, A. B. (1976). Turbulent drag reduction in dilute fiber suspensions: Mechanistic consideration. *AICHE Journal, 22*(4), 669–674.
Kato, H., Watanabe, K., & Ueda, K. (1972). Frictional resistance of rotating disk in dilute polymer solutions: Part I. enclosed disk. *Bulletin of the JSME, 15*(88), 1185–1196.
Kawada, H., & Tagori, T. (1973). *Proc. Ann. Meeting JSME.*
Kenis, P. R. (1968a). Drag reduction by bacterial metabolites. *Nature, 217*(5132), 940–942.
Kenis, P. R. (1968b). Effects of pH on the production of bacterial extracellular drag-reducing polymers. *Journal of Applied Microbiology, 16*(8), 1253–1254.
Kenis, P. R. (1969). Turbulent-flow drag reduction by polymers from marine and fresh water bacteria. *International Shipbuilding Progress, 16*(183), 342–348.
Kenis, P. R. (1971). Turbulent flow friction reduction effectiveness and hydrodynamic degradation of polysaccharides and synthetic polymers. *Applied Polymer Science, 15*, 607.
Kenis, P. R., & Hoyt, J. W. (1971). *Friction reduction by Algal and bacterial polymers* (Report No. NUC-TP-240). Naval Undersea Research & Development Center, San Diego.
Kerekes, R. J., & Douglas, W. J. (1972). Viscosity properties of suspensions at the limiting conditions for turbulent drag reduction. *The Canadian Journal of Chemical Engineering, 50* (2), 228–231.
Killen, J. M. (1972). *University of minnesota, St. Anthony Falls, hydraulic lab* (Project Report No. 123).
Killen, J. M., & Almo, J. (1969). An experimental study of the effects of dilute solutions of polymer additives on boundary layer characteristics. In C. S. Wells (Ed.), *Viscous drag reduction* (pp. 447–461). New York: Plenum Press.
Killian, F. P. (1970). Uber die verminderung des reibungswertes von grenzschichtstrosungen viscoelastischer flussigkeiten. In *Mitteilungen der versuchanstalt fur wasserbau and schiffsbau* (pp. Heft 51–Heft 52). Berlin, Germany.
Kim, O. K., Little, R. C., & Ting, R. Y. (1973). Polymer structural effects in turbulent drag reduction. *AICHE Chemical Engineering Progress Symposium Series No. 130, 69*, 39.

Kinnier, J. W. (1965). *A correlation between friction reduction and molecular size for the flow of dilute aqueous polyethyleneoxide solutions in pipes* (MS Thesis), US Naval Postgraduate School, Monterey.

Kirdyashkin, A. G. (1977). Fluid mechanics. *Soviet Research, 6*, 79.

Kobets, G. F. (1969). The mechanism of the influence of dissolved macromolecules (carboxymethylcellulose, polyvinyl alcohol, polysaccharides, slime of fish) on turbulent friction. *Bionika, 3*, 72.

Kobets, G. F., Zar'yalova, V. S., & Komarova, M. L. (1969). The influence of the slime of fish on turbulent friction. *Bionika, 3*, 80.

Kotenko, M., Oskarsson, H., Bojesen, C., & Nielsen, M. P. (2019). An experimental study of the drag reducing surfactant for district heating and cooling. *Energy, 178*, 72–78.

Kowalski, T. (1966). Reduction of frictional drag by non-newtonian additives. *Naval Engineers Journal, 78*(2), 293–297.

Kowalski, T. (1968a). Higher ships speeds due to injection of non-newtonian additives. Paper presented in Montreal on January 9th at the Eastern Canadian Section.

Kowalski, T. (1968b). Turbulence suppression and viscous drag reduction by non-newtonian additives. *Transactions RINA, 110*, 207–219.

Kruyt, H. R. (1949). In H. L. Booij. Association colloids, colloid science II, Elsevier Publishing Company, Amsterdam, p. 681.

Kumar, S. M., & Sylvester, N. D. (1973). Effects of a drag-reducing polymer on the turbulent boundary layer. *AICHE Chemical Engineering Progress Symposium Series No. 130, 69*, 1.

Kuo, Y., & Tanner, R. I. (1972). A burgers-type model of turbulent decay in non-newtonian fluid. *Transactions ASME Journal of Applied Mechanics, 39*(3), 661–666.

Kuriyama, K. (1962). Temperature dependence of micellar molecular weight of non-ionic surfactant in the presence of various additives. *Kolloid-Z u. Z-Polymere, 180*(1), 55–64.

Lacey, P. M. (1974). Drag reduction by long-chain polymers. *Chemical Engineering Science, 29* (6), 1495–1496.

Landahl, M. T. (1972). *Drag reduction (and shear flow) by polymer addition* (Report AFOSR-TR-73-1200). MIT Cambridge/MA, Dept. Aeronaut. Astronaut.

Lang, T. G. (1969). The effect of drag reduction and other improvements on the design and performance of submerged vehicles. In C. S. Wells (Ed.), *Viscous drag reduction* (pp. 313–330). New York: Plenum Press.

Latto, B., & Czaban, J. (1974). On the performance of turbomachinery in the presence of aqueous polymer solutions. In *Proc. intern. Conf. On drag reduction*. Cambridge, UK.

Latto, B., & Shen, C. H. (1970). Effect of dilute polymer injection on external boundary layer phenomena. *The Canadian Journal of Chemical Engineering, 48*(1), 34–38.

Lee, K. C., & Zakin, J. L. (1973). Drag reduction in hydrocarbon-aluminum soap polymer systems. *AICHE Chemical Engineering Progress Symposium Series No. 130, 69*, 45.

Lee, W. K., Vaselaski, R. C., & Metzner, A. B. (1974). Turbulent drag reduction in polymer solutions containing suspended fibers. *AICHE Journal, 20*(1), 128–133.

Lehmann, A. F., & Suessmann, R. T. (1972). *An experimental study of lift and drag of a hydrofoil with polymer ejection* (Report No. 72-94). Oceanics Inc.

Levy, J., & Davies, S. (1967). Drag measurements on a thin plate in dilute polymer solutions. *International Shipbuilding Progress, 14*(152), 166–175.

Liaw, G. C., Zakin, J. L., & Patterson, G. K. (1971). Effects of molecular characteristics of polymers on drag reduction. *AICHE Journal, 17*(2), 391–397.

Little, R. C. (1967). *Drag reduction by dilute polymer solutions in turbulent flow* (Report 6542). Naval Research Lab.

Little, R. C. (1969). Displacement of aqueous drag-reducing polymer solutions. *Industrial & Engineering Chemistry Fundamentals, 8*(3), 520–521.

Little, R. C. (1971). The effect of added salt on the flow of highly dilute solutions of poly(ethylene oxide) polymers. *Journal of Applied Polymer Science, 15*(12), 3117–3125.

Little, R. C., & Patterson, R. L. (1974). Turbulent friction reduction by aqueous poly(ethylene oxide) polymer solutions as a function of salt concentration. *Journal of Applied Polymer Science, 18*(5), 1529–1539.

Little, R. C., & Wiegard, M. (1971). The flow of very dilute polyox solutions into a region of sudden capillary tube enlargement. *Journal of Applied Polymer Science, 15*, 1515.

Little, R. C., Hansen, R. J., Hunston, D. L., Kim, O. K., Patterson, R. L., & Ting, R. Y. (1975). The drag reduction phenomenon. Observed characteristics, improved agents and proposed mechanisms. *Industrial & Engineering Chemistry Fundamentals, 14*(4), 283–296.

Lockett, F. J. (1964). Fluid dynamics approach to the Toms effect. *Nature, 222*, 937–939.

Lumley, J. L. (1964). Turbulence in non-newtonian fluids. *Physics of Fluids, 7*(3), 335–337.

Lumley, J. L. (1967). The Toms phenomenon: Anamolous effects in turbulent flows of dilute solutions of high molecular weight linear polymers. *Applied Mechanics Reviews, 20*(12), 1139–1149.

Lumley, J. L. (1969). Drag reduction by additives. *Annual Reviews of Fluid Mechanics, 1*, 367–384.

Lumley, J. L. (1970). *Concerning the behavior of dilute solutions of linear polymers*. In *Proc. drag reduction workshop*. Boston, MA: ONR.

Lumley, J. L. (1973). Drag reduction in turbulent flow by polymer additives. *Journal of Polymer Science Macromolecular Reviews, 7*(1), 263–290.

Lummus, J. L., & Randall, B. V. (1964). *Development of drilling fluid friction additives for project MOHOLE* (Report F64-P-54). Pan American Petroleum Corp. Research Dept. Job No. 3918.

Marris, A. W., & Wang, J. T.-S. (1965). In J. W. Hoyt (Ed.), *Symposium on rheology*. New York: ASME.

Maude, A. D., & Whitmore, R. L. (1958). The turbulent flow of suspensions in tubes. *Chemical Engineering Research and Design, 36a*, 297–305.

Maxson, A., Watson, L., Karandikar, P., & Zakin, J. (2017). Heat transfer enhancement in turbulent drag reducing surfactant solutions by agitated heat exchangers. *International Journal of Heat and Mass Transfer, 109*, 1044–1051.

McMillan, M. L. (1970). *Drag reduction and light scattering studies of aluminum disoaps in toulene* (PhD Thesis). Ohio State University, Columbus, OH.

McMillan, M. L., Hershey, H. C., & Baxter, R. (1971). Effects of aging, concentration, temperature, method of preparation, and other variables on the drag reduction of aluminum disoaps in toulene. *AICHE Chemical Engineering Progress Symposium Series No. 111, 67*, 27.

Mejean, L., & Boulos, M. I. (1976). Caracteristiques rheologiques des suspensions de tourbe. *The Canadian Journal of Chemical Engineering, 54*(5), 382–391.

Melton, L. L., & Malone, W. T. (1974). Fluid mechanics research and engineering applications in non-newtonian fluid systems. *SPE Journal, 4*(1), 56–66.

Mel'tser, L. Z., El'perin, I. T., Leventhal, L. I., & Kovalenko, V. S. (1972). Effect of polyacrylamide additive on hydraulic resistances of brine systems. *Kholod Tekh Tekhnol, 15*, 36.

Merkulov, V. I., & Khotinskaya, V. D. (1969). The mechanism of drag reduction in different types of fish (negative viscosity of dilute solutions of slime measured by the Oswald viscometer). *Bionika, 3*, 96.

Merrill, E. W., Mickley, H. S., & Ram, A. (1962). Instability in couette flow of solutions of macromolecules. *Journal of Fluid Mechanics, 13*, 86–90.

Meter, D. M. (1964). Tube flow of non-newtonian polymer solutions: Part II – turbulent flow. *AICHE Journal, 10*(6), 881–884.

Metzner, A. B., & Metzner, A. P. (1970). Stress levels in rapid extensional flows of polymeric fluids. *Rheologica Acta, 9*(2), 174–181.

Metzner, A. B., & Park, M. G. (1964). Turbulent flow characteristics of viscoelastic fluids. *Journal of Fluid Mechanics, 20*(2), 291–303.

Mih, W., & Parker, J. (1967). Velocity profile measurements and a phenomenological description of turbulent fiber suspension pipe flow. *TAPPI, 50*(5), 237–246.

Mueller, H. G., & Klein, J. (1980). Mechanical shear degradation of a polymer solution by capillary flow. *Makromolecular Rapid Communication, 1*(1), 27–29.

Mysels, K. J. (1949). *Flow of thickened fluids.* US Patent No. 2,492,173 (Dec 27).
Mysels, K. J. (1971). Early experiments with viscous drag reduction. *AICHE Chemical Engineering Progress Symposium Series No. 111, 67*, 1017.
Myska, J., & Simeckova, M. (1983). The shape of micelles of a complex soap causing the Toms effect. *Colloid & Polymer Science, 261*(2), 171–175.
Nadolink, R. H. (1973). *Friction reduction in dilute solutions of polystyrene* (Technical Report 4422). Naval Universea Systems Center.
Nadolink, R. H., & Haigh, W. W. (1995). Bibliography on skin friction reduction with polymers and other boundary-layer additives. *Applied Mechanics Reviews, 48*(7), 351–460.
Nagarajan, R., Davies, G. S., & Venkateswarlu, D. (1974). Drag reduction characteristics of polyacrylamide additives. *The Chemical Engineering Journal, 7*(3), 249–252.
Nash, T. (1956a). Modification of the bulk mechanical properties of water by complex formation in dilute solution. *Nature, 177*(4516), 948.
Nash, T. (1956b). Conjugation with lone-pair electrons. II the adsorption of napthols by cationic micelles in dilute aqueous solution. *Journal of Applied Chemistry, 6*(12), 539–546.
Nash, T. (1958). The interaction of some naphalene derivatives with a cationic soap below the critical micelle concentration. *Journal of Colloid Science, 13*(2), 134–139.
Nokes, R. F., Greene, H. L., & Thomas, L. C. (1971). Ventricular myograph tracing during polyacrylamide perfusion. In *Proc. 24th annual conf. on engg. in medicine and biology*. Las Vegas.
Oldroyd, J. G. (1948). A suggested method of detecting wall effects on turbulent flow through tubes. In *Proc. 1st intern. congr. on rheology, II* (p. 130). North Holland, Amsterdam.
Oltmann, P. (1969). Verusche zur verminderung des reibungwiderstandes von schiffsmodelen. *Schiff und Hafen, 21*, 3.
Ousterhout, R. S., & Hall, C. D. (1961). Reduction of friction loss in fracturing operations. *The Journal of Petroleum Technology, 13*(3), 217–222.
Palyvos, J. A. (1974). *Drag reduction and associated phenomena in internal and external liquid flows* (Report No. 741). Thermodynamics and Transport Phenomena Lab, National Techn Univ., Athens 147, Greece.
Parker, C. A., & Joyce, T. A. (1974). Drag reduction and molecular structure. The interaction of polyethylene amine with some linear high polymers. *Journal of Applied Polymer Science, 18*(1), 155–165.
Paterson, R. W., & Abernathy, F. H. (1970). Turbulent flow drag reduction and degradation with dilute polymer solutions. *Journal of Fluid Mechanics, 43*(4), 689–710.
Patterson, G. K., & Zakin, J. L. (1968). Prediction of drag reduction with a viscoelastic model. *AICHE Journal, 14*(3), 434–439.
Patterson, G. K., Zakin, J. L., & Rodriguez, J. M. (1969). Drag reduction – polymer solutions, soap solutions and solid particle suspensions in pipe flow. *Industrial and Engineering Chemistry, 61* (1), 22–30.
Pereira, A. S., Mompean, G., Thais, L., & Soares, E. J. (2017). Transient aspects of drag reducing plane couette flows. *Journal of Non-Newtonian Fluid Mechanics, 241*, 60–69.
Peterlin, A. (1970). Molecular model of drag reduction by polymer solutes. *Nature, 227*(5258), 598–599.
Peyser, P. (1973). The drag reduction of chrysotile asbestos dispersions. *Journal of Applied Polymer Science, 17*(2), 421–431.
Peyser, P., & Little, R. C. (1971). The drag reduction of dilute polymer solutions as a function of solvent power, viscosity and temperature. *Journal of Applied Polymer Science, 15*(11), 2623–2637.
Pfenninger, W. (1967). *A hypothesis of the reduction of turbulent friction drag in fluid flows by means of additives.* Northrop Corp. Norair Division Report BLC-179.
Pilpel, N. (1954). On gel formation in soaps. *Journal of Colloid Science, 9*(4), 285–299.
Pilpel, N. (1966a). Viscoelasticity of aqueous soap solutions. Part 3. *Transactions of the Faraday Society, 62*, 1015–1022.

Pilpel, N. (1966b). Viscoelasticity of aqueous soap solutions. Part 4. – effect of alcohols. *Transactions of the Faraday Society, 62*, 2941–2952.

Pirih, R. J., & Swanson, W. M. (1972). Drag reduction and turbulence modification in rigid particle suspensions. *The Canadian Journal of Chemical Engineering, 50*(2), 221–227.

Polishchunk, A. M., Raiskii, Y. D., & Temchin, A. Z. (1972). Effect of small addition of polyisobutylene on the turbulent flow of kerosene in a pipe. *Neftyanoe Khozyaistvo (Petroleum Industry), 50*, 60.

Pollert, J. (1977). *Proc. second intern. conf. on drag reduction* (pp. B3–B37). Cambridge, UK: BHRA Fluid Engg.

Poreh, M., Zakin, J. L., Brosh, A., & Warsharsky, M. (1970). Drag reduction in hydraulic transport of solids. *Proceedings of the ASCE Hydraulics Division, 4*, 903–909.

Prather, R. J. (1966). *Investigations of the ultrasonic dynamic viscoelastic properties of aqueous poly(ethylene oxide) solutions* (MS Thesis). US Naval Postgraduate School, Monterey.

Procaccia, I., L'vov, V. S., & Benzi, R. (2008). Theory of drag reduction by polymers in wall-bounded turbulence. *Reviews of Modern Physics, 80*(1), 225–247.

Pruitt, G. T., & Crawford, H. R. (1963). *Drag reduction, rheology, and capillary end effects of some dilute polymer solutions.* Final Report, Westco Research on Contract 60530-8250 to Naval Ordinance Test Section.

Pruitt, G. T., & Crawford, H. R. (1965). *Effect of molecular weight and segmental constitution on the drag reduction of water soluble polymers* (Report No. DTMB-1).Western Co. under Contract No. Nonr 4306 (00).

Pruitt, G. T., Rosen, B., & Crawford, H. R. (1966). *Effect of polymer coiling on drag reduction.* Western Co. Report DTMB-2 Nonr 4306 (00).

Pruitt, G. T., Simmons, C. M., Neill, G. H., & Crawford, H. R. (1964). *A method to minimise costs of pumping fluids containing friction reducing additives.* SPE paper No 997.

Pyatetskii, V. F., & Savshenko, Y. N. (1969). The influence of slime on the flow resistance of fish. *Bionika, 3*, 90.

Radin, I., Zakin, J. L., & Patterson, G. K. (1969). Exploratory drag reduction studies in non-polar soap systems. In C. S. Wells (Ed.), *Viscous drag reduction* (pp. 213–231). New York: Plenum Press.

Radin, I., Zakin, J. L., & Patterson, G. K. (1973). Drag reduction of solid-liquid suspensions in pipe flow. *Nature Physical Sciences, 246*(149), 11–12.

Ram, A., & Kadim, A. (1970). Shear degradation of polymer solutions. *Journal of Applied Polymer Science, 14*(8), 2145–2156.

Ram, A., Finkelstein, F., & Elata, C. (1967). Reduction of friction in oil pipelines by polymer additives. *Industrial & Engineering Chemistry Process Design and Development, 6*(3), 309–313.

Ramakrishnan, C., & Rodriguez, F. (1973). Drag reduction in nonaqueous liquids. *AICHE, Chemical Engineering Progress Symposium Series No. 130, 69*, 52.

Ripkin, J. F., & Pilch, M. (1963). *Studies of the reduction of pipe friction with the non-newtonian additive CMC, St. Anthony Falls hydraulic lab.* Technical Paper No. 42, Series B.

Robertson, A. A., & Chang, M. V. (1967). Flocculation studies of fiber suspensions: Influence of zeta potential. *Pulp and Paper Magazine of Canada, 68*, T438.

Robertson, A. A., & Mason, S. G. (1957). The flow characteristics of dilute fiber suspensions. *TAPPI, 40*(5), 326–344.

Rodriguez, F. (1971). Drag reduction by a polymeric aluminum soap. *Nature Physical Sciences, 230*(15), 152–153.

Rodriguez, J. M., Zakin, J. L., & Patterson, G. K. (1967). Correlation of drag reduction with modified deborah number for dilute polymer solutions. *SPE Journal, 7*, 325–332.

Rosen, M. W., & Cornford, N. E. (1971). Fluid friction of fish slimes. *Nature, 234*(5323), 49–51.

Rubin, H. (1972). Drag reduction application in fire-fighting systems. *Proceedings of ASCE Journal Sanitary Engineering, 98*, 1.

Rubin, H., & Elata, C. (1966). Stability of couette flow of dilute polymer solutions. *Physics of Fluids, 9*(10), 1929–1933.

Rubin, H., & Elata, C. (1971). Turbulent flow of dilute polymer solutions through an annulus. *AICHE Journal, 17*(4), 990–996.

Ruckenstein, E. (1973). A note on the mechanism of drag reduction. *Journal of Applied Polymer Science, 17*(10), 3239–3240.

Rudd, M. J. (1971). Laser dopplermeter and polymer drag reduction. *AICHE Chemical Engineering Progress Symposium Series, 111*(67), 21.

Rudd, M. J. (1972). Velocity measurements made with a laser dopplermeter on the turbulent pipe flow of a dilute polymer solution. *Journal of Fluid Mechanics, 51*(4), 673–685.

Ruszczycky, M. A. (1965). Sphere drop tests in high polymer solutions. *Nature, 206*, 614–615.

Sanders, J. V., Henderson, B. H., & White, R. (1973). Effects of polyethylene oxide solutions on the performance of a small propeller. *Journal of Hydronautics, 7*(3), 124–128.

Sarpkaya, T. (1973). Lift and drag measurements on a hydrofoil in dilute polyox solutions. *Nature, 241*(5385), 114–115.

Savins, J. G. (1961). Some comments on pumping requirements for non-newtonian fluids. *Journal Institute of Petroleum, 47*, 329.

Savins, J. G. (1964). Drag reduction characteristics of solution of macromolecules in turbulent pipe flow. *SPE Journal, 4*(3), 203–214.

Savins, J. G. (1967). A stress controlled drag reduction phenomenon. *Rheologica Acta, 6*(4), 323–330.

Savins, J. G. (1968). *Method of decreasing friction loss in turbulent liquids.* US Patent No. 3,361,213.

Savins, J. G. (1969). Contrasts in the solution drag reduction characteristics of polymer solutions and micellar systems. In C. S. Wells (Ed.), *Viscous drag reduction* (pp. 183–212). New York: Plenum Press.

Scott, D. (1969). Slippery water in fire hoses. *Science, 164*(3887), 1466.

Sellin, R. H. (1978). Drag reduction in sewers: First results from a permanent installation. *Journal of Hydraulic Research, 16*(4), 357–371.

Sellin, R. H., & Barnard, B. J. (1970). Open channel applications for dilute polymer solutions. *Journal of Hydraulic Research, 8*(2), 219–228.

Sellin, R. H., & Barnard, B. J. (1971). The use of friction reducing additives to increase the capacity of storm water sewers. *Journal of the Institution of Municipal Engineers, 98*, 207.

Sellin, R. H., & Ollis, M. J. (1980). Polymer drag reduction in large pipes and sewers: Results of recent field trials. *Journal of Rheology, 24*(5), 667–684.

Sellin, R. H., Hoyt, J. W., & Scrivener, O. (1982). The effect of drag reducing additives on fluid flows and their industrial applications. Part 1: Basic aspects. *Journal of Hydraulic Research, 20* (1), 29–68.

Sellin, R. H., Hoyt, J. W., Pollert, J., & Scrivener, O. (1982). The effect of drag reducing additives on fluid flows and their industrial applications. Part 2: Present applications and future proposals. *Journal of Hydraulic Research, 20*, 235–292.

Seyer, F. A., & Metzner, A. B. (1967a). Turbulent flow properties of viscoelastic fluids. *The Canadian Journal of Chemical Engineering, 45*(3), 121–126.

Seyer, F. A., & Metzner, A. B. (1967b). *Turbulence phenomena in drag reducing systems, 60th annual AICHE meeting*, New York.

Shaver, R. G., & Merrill, E. W. (1959). Turbulent flow of pseudoplastic polymer solutions in straight cylindrical tubes. *AICHE Journal, 5*(2), 181–188.

Sheffer, H. (1948). Aluminum soaps as high polymers. *Canadian Journal of Research, 26b*(6), 481–498.

Shenoy, A. V. (1976). Drag reduction with surfactants at elevated temperatures. *Rheologica Acta, 15*(11–12), 658–664.

Shenoy, A. V. (1984). A review on drag reduction with special reference to micellar systems. *Colloid & Polymer Science, 262*(4), 319–337.

Shi, H., Wang, Y., Fang, B., Talmon, Y., Ge, W., Raghavan, S. R., et al. (2011). Light-responsive threadlike micelles as drag reducing fluids with enhanced heat-transfer capabilities. *Langmuir, 27*(10), 5806–5813.

Shin, H. (1965). *Reduction of drag in turbulence by dilute polymer solutions* (ScD Thesis). MIT, Cambridge, MA.

Song, C. S., & Tsai, F. Y. (1966). *Experimental investigation of taylor instability using non-newtonian fluids.* Project Report Mp-84, St., Anthony Falls Hydraulic Lab.

Sylvester, N. D., & Smith, P. S. (1979). The concentration and friction velocity effects on drag reduction by Dowell-APE in kerosene. *Industrial & Engineering Chemistry Product Research and Development, 18*(1), 47–49.

Thomas, D. G. (1962). Transport characteristics of suspensions: Part IV. Friction loss of concentrated – flocculated suspensions in turbulent flow. *AICHE Journal, 8*(2), 266–271.

Thorne, P. F. (1974). Drag reduction in fire fighting. In *Proc. intern. conf. on drag reduction*. Cambridge, UK.

Thorne, P. F., Theobald, C. R., & Mahendran, P. (1975). Drag reduction in fire hose trials at fire service technical college, Part 1. Experiments and Results, Research Note 1033, Part 2: Analysis and Application of results, Research Note 1043.

Ting, R. Y. (1982). Some molecular effects in drag reduction: A summary. *Chemical Engineering Communications, 15*(5–6), 331–342.

Ting, R. Y., & Kim, O. K. (1973). Water-soluble polymers. In N. M. Bikales (Ed.), *Drag reduction properties of high molecular weight polyacrylamide and related polymers*. New York: Plenum Press.

Tomita, Y. (1970). Pipe flows of dilute aqueous polymer solution: Part 1, experimental study of pipe friction coefficient. *Bulletin of JSME, 13*(61), 926–934.

Toms, B. A. (1948). Some observations on the flow of linear polymer solutions through straight tubes at large Reynolds numbers. In *Proc. 1st intern. congr. on rheology, II* (pp. 135–141). Scheveningen, North Holland, Amsterdam.

Tothill, J. T. (1967). Ships in restricted channels – a correlation of model tests, field measurements, and theory. *Marine Technologies, 4*, 111.

Treiber, K. L., & Sieracki, L. M. (1970). *The effect of non-newtonian friction reducing additives in a diesel fuel pipeline* (Report No. 101-Z). Columbia Res. Corp.

Tulin, M. P. (1966). Hydrodynamic aspects of macromolecular solutions. In *Proc. 6th symp. on naval hydrodynamics ACR-136* (p. 3). ONR: Washington, DC.

Van Driest, E. R. (1970). Turbulent drag reduction of polymeric solutions (long chain molecule additive effect on drag reduction in turbulent flow of aqueous polymeric solutions). *Journal of Hydronautics, 4*(3), 120–126.

Van Driest, E. R. (1971). Turbulent flow of non-Newtonian fluids. In *Proc. 9th intern. symp. on space tech. and science*. Tokyo.

Vanasse, R., Coupal, B., & Boulos, M. I. (1979). Hydraulic transport of peat moss suspensions. *The Canadian Journal of Chemical Engineering, 57*(2), 238–241.

Vanoni, V. A. (1946). Transportation of suspended sediment by water. *Transactions ASCE, 111*, 67–133.

Vanoni, V. A., & Nomicos, G. N. (1960). Resistance properties of sediment laden streams. *Transactions ASCE, 125*(1), 1140–1167.

Varshney, A., & Steinberg, V. (2018). Drag enhancement and drag reduction in viscoelastic flow. *Physics Review Fluids, 3*, 103302.

Vaselaski, R. C., & Metzner, A. B. (1974). Drag reduction in the turbulent flow of fiber suspensions. *AICHE Journal, 20*(2), 301–306.

Virk, P. S. (1966). *The toms phenomenon – Turbulent pipe flow of dilute polymer solutions*, Mass. Inst. of Tech., (ScD Thesis).

Virk, P. S. (1971a). Drag reduction in rough pipes. *Journal of Fluid Mechanics, 45*(2), 225–246.

Virk, P. S. (1971b). An elastic sublayer model for drag reduction by dilute solutions of linear macromolecules. *Journal of Fluid Mechanics, 45*(3), 417–440.

Virk, P. S. (1975). Drag reduction fundamentals. *AICHE Journal, 21*(4), 625–656.

Virk, P. S., Mickley, H. S., & Smith, K. A. (1970). The ultimate asymptote and mean flow structure in Toms' phenomenon. *Transactions ASME Journal of Applied Mechanics, 37*(2), 488–493.

Vogel, V. M., & Patterson, A. M. (1964). An experimental investigation of the effects of additives injected into the boundary layer of an underwater body. In *Proc. 5th symp. on naval hydrodynamics ONR-ACR-112* (p. 975). Berge.

Walsh, M. (1967a). *On the turbulent flow of dilute polymer solutions* (PhD Thesis). California Institute of Technology.

Walsh, M. (1967b). Theory of drag reduction in dilute high-polymer flows. *International Shipbuilding Progress, 14*(152), 134–139.

Walters, K., Barnes, H. A., & Dodson, A. C. (1971). Drag reduction in unconventional flow situations. *AICHE Chemical Engineering Progress Symposium Series No. 111, 67*, 1.

Walters, R. R., & Wells, C. S. (1971). An experimental study of turbulent diffusion of drag-reducing polymer additives. *Journal of Hydronautics, 5*(2), 65–72.

Wang, Y., Yu, B., Zakin, J. L., & Shi, H. (2011). Review on drag reduction and its heat transfer by additives. *Advances in Mechanical Engineering, 17*. https://doi.org/10.1155/2011/478749

Warholic, M. D., Massah, H., & Hanratty, T. J. (1999). Influence of drag-reducing polymers on turbulence: Effects of Reynolds number, concentration and mixing. *Experiments in Fluids, 27* (5), 461–472.

Wei, T., & Willmarth, W. W. (1992). Modifying turbulent structure with drag-reducing polymer additives in turbulent channel flows. *Journal of Fluid Mechanics, 245*, 619–641.

Wells, C. S. (1969). An analysis of uniform injection of a drag reducing fluid into a turbulent boundary layer. In C. S. Wells (Ed.), *Viscous drag reduction* (pp. 361–382). New York: Plenum Press.

White, A. (1966a). Effect of polymer additives on boundary layer separation and drag of submerged bodies. *Nature, 211*(5056), 1390.

White, A. (1966b). Turbulent drag reduction with polymer additives. *Journal of Mechanical Engineering Science, 8*(4), 452–455.

White, A. (1967a). Flow characteristics of complex soap systems. *Nature, 214*(5088), 585–586.

White, A. (1967b). Drag of spheres in dilute high polymer solutions. *Nature, 216*(5119), 994–995.

White, A. (1968). Studies of flow characteristics of dilute high polymer solutions. *Henderson College of Technology Research Bulletin*, (5), 113.

White, A., & Hemmings, J. A. (1976). *Drag reduction bt additives: Review and bibliography*. Cranfield, UK: BHRA Fluid Engg.

White, C. M., & Mungal, M. G. (2008). Mechanics and prediction of turbulent drag reducation with polymer additives. *Annual Review of Fluid Mechanics, 40*, 235–256.

Whitsitt, N. F., Harrington, J. J., & Crawford, H. R. (1968). *Effect of wall shear stress on drag reduction of viscoelastic fluids*. Western Co. Report No. DTMB-3, Contract No. Nonr-4306(00).

Williams, M. C. (1965). Normal stress and viscosity measurements for polymer solutions in steady cone-and-plate shear. *AICHE Journal, 11*(3), 467–473.

Wolff, J. H., & Cahn, R. D. (1971). *Lifting surfaces in polymer solutions* (p. 3653). Bethesda, MD: Naval Ship Research Development Center Report.

Wu, J. (1969). Drag reduction in external flows of additive solutions. In C. S. Wells (Ed.), *Viscous drag reduction* (pp. 331–350). New York: Plenum Press.

Wu, J. (1971). Surface containing of polymer solution and pulsative ejection. *Nature Physical Sciences, 231*(24), 150.

Wu, J. (1973). Injection of drag-reducing polymers into a turbulent boundary layer. *Journal of Hydronautics, 7*(3), 129–132.

Zakin, J. L. (1972). Effects of age and water content on drag reduction in aluminum disoap-hydrocarbon solutions. *Nature Physical Sciences, 235*(57), 97–98.

Zakin, J. L., & Chang, J. L. (1972). Nonionic surfactants as drag reducing additives. *Nature Physical Sciences, 239*(89), 26–29.

Zakin, J. L., & Chang, J. L. (1974). Polyoxyethylene alcohol non-ionic surfactants as drag reducing additives. In *Proc. intern. conf. on drag reduction*. Cambridge, UK.

Zakin, J. L., & Hunston, D. L. (1980). Effect of polymer molecular variables on drag reduction. *Journal of Macromolecular Science Physics, B18*(4), 795–815.

Zakin, J. L., Lu, B., & Bewersdorff, H. W. (1998). Surfactant drag reduction. *Reviews in Chemical Engineering, 14*(4–5), 255–320.

Zakin, J. L., Poreh, M., Brosh, A., & Warsharsky, M. (1971). Exploratory study of friction reduction in slurry flows. *AICHE Chemical Engineering Progress Symposium Series No. 111, 67*, 85–89.

Zandi, I. (1967). Decreased head losses in raw water conduits. *Journal of American Water Works Association, 59*(2), 213–226.

Zhu, L., Bai, X., Krushelnycky, E., & Xi, L. (2019). Transient dynamics of turbulence growth and bursting: Effects of drag-reducing polymers. *Journal of Non-Newtonian Fluid Mechanics, 266*, 127–142.

Chapter 2
Velocity Profiles and Friction Factors in Turbulent Pipe Flows

Expressions for turbulent velocity profiles and friction factors during flow of drag reducing fluids have been developed using similar assumptions like those for Newtonian fluids and power-law fluids. The overall cross section of the pipe is assumed to be divisible in two regions:

1. A turbulent core, wherein inertial effects are dominant
2. A wall region, wherein the constitutive properties of the fluid assume importance

The laminar sublayer in the wall region is assumed to extend from the wall ($y = 0$) to $y = y_{SL}$ and the turbulent core from $y = y_{SL}$ to $y = R$.

Smooth Straight Circular Pipes

Several empirical relationships for correlating and predicting drag reduction effects have been developed (Kilbane & Greenkorn, 1966; Meter, 1964; Seyer & Metzner, 1967a; Virk, Merrill, Mickley, Smith, & Mollo-Christensen, 1967). There are correlations for drag reducing fluids, which have in part a theoretical reasoning (Astarita, Greco, & Nicodemo, 1969; Meyer, 1966; Seyer & Metzner, 1969a).

The most popular approach (Seyer & Metzner, 1969a) uses the logarithmic similarity laws for drag reducing systems based on theoretical arguments similar to those for Newtonian fluids (Millikan, 1939) and for purely viscous non-Newtonian fluids (Dodge & Metzner, 1959).

Fully Developed Velocity Profiles and Friction Factors

The time-averaged velocity at a point during turbulent flow of a drag reducing fluid would depend on six independent variables, namely, R, ρ, τ_w, μ, θ_{fl}, and y. These are

A. Shenoy, *Rheology of Drag Reducing Fluids*,
https://doi.org/10.1007/978-3-030-40045-3_2

no different from those appearing for Newtonian fluids except for the additional parameter θ_{fl} to include the elasticity of drag reducing fluids. Based on the two-region model mentioned above, it is obvious that in the wall region

$$u = \Phi_1(\rho, \tau_{\mathrm{w}}, \mu, \theta_{\mathrm{fl}}, y) \tag{2.1}$$

and in the turbulent core

$$u_{\mathrm{m}} = u = \Phi_2(R, \rho, \tau_{\mathrm{w}}, y) \tag{2.2}$$

Following the conventional dimensional analysis (Langhaar, 1951) used for drag reducing fluids (Seyer & Metzner, 1969b), it can be shown that over the entire cross section

$$\frac{u}{u^*} = \Phi(\mathrm{Re}^*, \xi_{\mathrm{R}}, \mathrm{De}) \tag{2.3}$$

where u^* is the friction or shear velocity defined as $\sqrt{\tau_{\mathrm{w}}/\rho}$, Re^* is the Reynolds number based on friction velocity and defined as $R\,u^*\rho/\mu$, and ξ_{R} is the dimensionless location parameter defined as y/R.

Expressions for the velocity profiles during turbulent flow have been developed by many investigators (Elata, Lehrer, & Kahanovitz, 1966; Meyer, 1966; Seyer & Metzner, 1969a; Virk, 1975), and though each has proposed a different expression for the turbulent core region, there is an agreement regarding the expression in the wall region given as

$$u^+ = y^+ \text{ for } y \le y_{\mathrm{SL}} \tag{2.4}$$

where u^+ is a dimensionless velocity defined as u/u^* and y is the distance-based Reynolds number defined as $yu^*\delta/\mu$.

For the turbulent core, it was observed (Meyer, 1966) that the velocity profiles in turbulent pipe flow of very dilute drag reducing fluids could be correlated by a modified form of the universal law of the wall as follows:

$$u^+ = 2.5\ \ln y^+ + 5.5 + \overline{\alpha}^* \ln\left(\frac{u^*}{u^*_{\mathrm{crit}}}\right) \tag{2.5}$$

The parameters $\overline{\alpha}^*$ and u^*_{crit} are, respectively, measures of the drag reducing effectiveness and onset of the wall shear stress above which drag reduction occurs, and these are constant for a given fluid system. The effect of these parameters on the velocity profile is to create a shift in the logarithmic portion without a change in slope from the Newtonian law of the wall – thus indicating an effective thickening of the viscous sublayer with no apparent change in the mixing-length distribution.

The modified form of the conventional Newtonian law-of-the-wall expression (Elata et al., 1966) gave a different type of correction term involving the Deborah number as follows:

$$u^+ = 2.5\ \ln y^+ + 5.5 + \overline{\alpha}^{**}\ln\text{De} \tag{2.6}$$

A more general equation of the velocity profile for the turbulent core in smooth circular pipes (Seyer & Metzner, 1969a) has the following form:

$$u^+ = A'\ \ln y^+ + B' \tag{2.7}$$

where

$$A' = 2.46 \tag{2.8}$$

$$B' = 5.6 + 1.55\ \text{De} \quad (\text{for } 0 \leq \text{De} \leq 10) \tag{2.9}$$

De is of the form given below:

$$\text{De} = \frac{\theta_{\text{fl}} u^{*2}}{\upsilon} \tag{2.10}$$

Since drag reducing fluids are known to be Newtonian in viscosity but exhibit mild elasticity, use is made of a dimensionless Deborah number (De) defined as the ratio of the fluid relaxation time (θ_{fl}) and characteristic process time (ν/u^{*2}). This is the definition that will be used throughout herein, though one might find slight variations in the definitions used by different researchers, and hence caution needs to be exercised when applying it to the works of other researchers.

During steady shear flow of polymer solutions, molecules are known to undergo extensions. However, due to the elastic nature of the fluids, the molecules show a tendency to relax. The material constant θ_{fl} can be considered as time required for molecular relaxation and would therefore be a measure of the degree of fluid elasticity.

There is considerable discussion in the literature (Astarita, 1965; Seyer & Metzner, 1969b; Virk, 1975) wherein it has been explicitly demonstrated that choice of the characteristic time scales of the above kind for defining De is completely adequate for correlating the frictional characteristics in drag reducing flows. Reported experimental studies on the determination of fluid relaxation times (θ_{fl}) published in the literature (Seyer & Metzner, 1969b) show that (θ_{fl}) varies as $\dot{\gamma}^{-m}$, where m lies between 0.5 and 1.0. The general practice is to assume m as equal to 1 so that Deborah number can be taken as a constant independent of shear rate knowing that (u^{*2}/ν) is proportional to wall shear rate.

Another expression for the velocity profile is based on the elastic sublayer model (Virk, 1975). The essential physical notion is that the stimulation of polymer molecules by a turbulent shear flow creates a zone, called the elastic sublayer, which is characteristic of the drag reduction phenomenon. The elastic sublayer originates at the onset of drag reduction and grows with increasing drag reduction until it finally occupies the entire pipe cross section at maximum drag reduction. The mean velocity profiles during drag reduction would thus show two extremes: at low drag reduction, the entire outer flow, $y^+ > 50$, say, is shifted upward from parallel to the Newtonian wall law by an amount S^+, defined as the effective slip, whereas, at maximum drag reduction, the ultimate velocity profile has a semilogarithmic form like the Newtonian law of wall except for the mixing-length constant, which equals 0.085 instead of 0.4 as in the Newtonian case. Thus, the following expressions were proposed (Virk, 1975):

$$u^+ = 11.7 \ln y^+ - 17.0 \quad \text{for } y^+_{\text{SL}} < y^+ < y^+_{\text{E}} \tag{2.11}$$

In the Newtonian plug region, the profile is parallel-shifted upward, relative to the Newtonian law of wall as

$$u^+ = 2.5 \ln y^+ + 5.5 + 9.2 \ln \frac{y^+_{\text{E}}}{y^+_{\text{SL}}} \quad \text{for } y^+_{\text{E}} \le y^+ \le R \tag{2.12}$$

A schematic diagram of the velocity distribution as described above is shown in Fig. 2.1 (Sellin, Hoyt, & Scrivener, 1982).

From the velocity profile given by Eq. (2.7), the relationship between friction factor and Reynolds number could be easily derived by determining the average velocity by integration and assuming a linear velocity profile in the viscous sublayer to yield

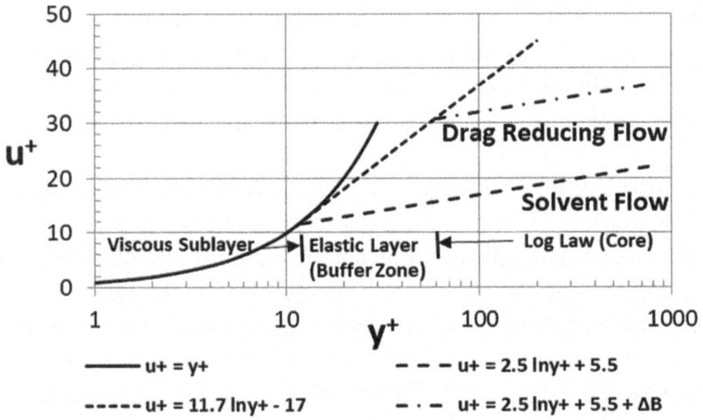

Fig. 2.1 Velocity distribution based on typical velocity profile models for turbulent flow of drag reducing fluids (Sellin, Hoyt, & Scrivener, 1982). (Reprinted with permission from Taylor & Francis and Copyright Clearance Center through RightsLink)

$$\sqrt{\frac{2}{f}} = A'(1-\xi_1)^2 \ln \frac{\text{Re}\sqrt{f}}{2\sqrt{2}} + B'(1-\xi_1)^2 - G \tag{2.13}$$

Equation (2.13) was found (Seyer & Metzner, 1969a) to fit available friction factor data for drag reducing fluids within a mean deviation of 4.2% for a value of $G = 3.0$. Despite the excellent agreement between the correlating equations and experimental data, the velocity profile given by Eq. (2.7) has an inherent incongruity in failing to predict a zero-velocity gradient at the centerline. The logarithmic expressions for the velocity profile in the turbulent core as given by Eqs. (2.6), (2.7), (2.11), and (2.12) can predict the velocity distributions even beyond the range of experimentation due to their asymptotic nature. However, these expressions all have an incongruity in that they fail to predict a zero-velocity gradient at the center of the pipe. Such an incongruity has existed in all the earlier suggested velocity profiles both for Newtonian fluids (Nikuradse, 1932) and inelastic non-Newtonian fluids (Bogue & Metzner, 1963; Clapp, 1961; Dodge & Metzner, 1959).

Velocity profile models for turbulent flow in smooth pipes were suggested for Newtonian fluids (Stein, Kessler, & Greenkorn, 1980) and for inelastic non-Newtonian power law fluids (Shenoy & Saini, 1982) by proper adjustment of the parameters so that the centerline velocity gradient was zero. The same approach was extended (Shenoy & Talathi, 1985) to mildly elastic drag reducing fluids so that a velocity profile, devoid of this limitation of the centerline velocity not being zero, could be evolved. A model for mean velocity profile of turbulent water flow with added drag reducing surfactants based on modified Prandtl's mixing length hypothesis was introduced (Krope, Krope, & Lipus, 2005) and included three parameters, which were determined numerically from experimental datasets. The minimum transport velocity model was used (Ramadan, Saasen, & Skalle, 2004) for drag reducing polymers for predicting the minimum fluid velocity needed to initiate motion of sand bed particles.

The velocity profile model proposed by Shenoy and Talathi is devoid of the limitation of the centerline velocity not being zero and is, therefore, the correct wall theoretical velocity profile expression for the turbulent core during flow of drag reducing fluids (Shenoy & Talathi, 1985). The form of expression assumed is similar to that proposed earlier (Seyer & Metzner, 1969a), except for an added correction term $C(\xi, \text{De})$.

It is thus assumed that the velocity profile in the turbulent core is of the following form:

$$u^+ = A[\ln y^+ + C(\xi, \text{De})] + B(\text{De}) \tag{2.14}$$

where $C(\xi, \text{De})$ is a correction function having the following form similar to that used before (Shenoy & Saini, 1982; Stein et al., 1980), based on the suggestion given earlier (Hinze, 1955):

$$C(\xi, \mathrm{De}) = \sigma_1(\mathrm{De}) \exp\left\{ -\frac{1}{2}\left[\frac{\xi - 0.8}{\sigma_2(\mathrm{De})}\right]^2 \right\} \tag{2.15}$$

In order that Eq. (2.14) should satisfy the condition that the velocity gradient becomes zero at the centerline and also give the form of the friction factor given by Eq. (2.13) for drag reducing fluids when integrated, it is imperative that the correction function should satisfy the following condition:

$$C'(1, \mathrm{De}) = -\frac{0.2\sigma_1(\mathrm{De})}{\sigma_2{}^2(\mathrm{De})} \exp\left\{ -\frac{0.02}{\sigma_2{}^2(\mathrm{De})} \right\} = -1 \tag{2.16}$$

The following forms of $\sigma_1(\mathrm{De})$ and $\sigma_2(\mathrm{De})$ are seen to be adequate for all values of De from 0 to 10 to satisfy the above condition:

$$\sigma_1(\mathrm{De}) = 0.4398 + 0.123\ \mathrm{De} + 0.0135\ \mathrm{De}^2 \tag{2.17}$$

$$\sigma_2(\mathrm{De}) = 0.254\ (1 + 0.2\ \mathrm{De}) \tag{2.18}$$

Note that for De $=$ 0, the values of $\sigma_1(0)$ and $\sigma_2(0)$ are equivalent to those obtained for Newtonian fluids (Stein et al., 1980).

Equation (2.14) is to be integrated over the entire cross section of the turbulent core, i.e., from the laminar sublayer to the centerline. Since the integration involves evaluation of complex error functions whose definitive values must be determined, some of the factors are evaluated over the entire cross section of the pipe as a first approximation. Using the definition

$$y^+ = \xi \frac{\mathrm{Re}\sqrt{f}}{2\sqrt{2}} \tag{2.19}$$

an implicit expression for the friction factor can be derived in a manner like that for the Newtonian case (Stein et al., 1980) to yield the following:

$$\sqrt{\frac{2}{f}} = A(1 - \xi_1)^2 \ln \frac{\mathrm{Re}\sqrt{f}}{2\sqrt{2}} + (1 - \xi_1)^2 B - AD \tag{2.20}$$

It was found that the values of D, obtained during the determination of the mean dimensionless velocity, incorporated terms obtained from the evaluation of the error functions and could be approximated by the following equation:

$$D = 1.3676\ \left(1 - 0.09\ \mathrm{De} - 0.01\ \mathrm{De}^2\right) \tag{2.21}$$

Comparison of Eqs. (2.13) and (2.20) yields the expressions for A and B for the first approximation of $\xi_1 = 0$, which, when substituted in Eq. (2.14) along with the expression for C, gives the dimensionless velocity profile for the turbulent flow of mildly elastic drag reducing fluids as

$$u^{+} = 2.46\left[\ln y^{+} + \left(0.4398 + 0.123\ \mathrm{De} + 0.0135\ \mathrm{De}^{2}\right)\right.$$
$$\left.\times \exp\left\{-\frac{(\xi - 0.8)^{2}}{0.129(1 + 0.2\mathrm{De})^{2}}\right\} + 1.3676\ \left(1 - 0.09\mathrm{De} - 0.01\ \mathrm{De}^{2}\right)\right]$$
$$+ 5.6 + 1.55\mathrm{De} - G \tag{2.22}$$

In Eq. (2.22), if G is taken equal to 3.0, as suggested (Seyer & Metzner, 1969a), then an error of about 4% is seen during the comparison of Eqs. (2.7) and (2.22). However, a choice of $G = 4.0$ yields an excellent agreement within less than 1.0% error for all values of De over a range of Reynolds numbers from 10^4 to 10^6. Furthermore, the model predicts a zero-velocity gradient at the centerline for all De values ($0 \leq \mathrm{De} < 10$), while the expression in Eqs. (2.7) (Seyer & Metzner, 1969a) fails in this respect.

The value of ξ in Eq. (2.22) can be obtained by rearrangement of the terms in the definition of y^+ given in Eq. (2.19) as

$$\xi = \frac{y^{+} 2\sqrt{2}}{\mathrm{Re}\ \sqrt{f}} \tag{2.23}$$

The value of f is often estimated from the expression given (Seyer & Metzner, 1969a) for drag reducing fluids over a wide range of Reynolds numbers from 4×10^3 to 10^6 as given in Eq. (2.13). Here ξ_1 is the dimensionless sublayer thickness defined by Eq. (2.23) with y^+ replaced by y^{+}_{SL}.

Due to the implicit nature of Eq. (2.13) in terms of f, an iterative procedure becomes necessary to get estimates of f. However, the following explicit expression (Shenoy, 1988) is found to give values of f within an error boundary of ±5% for the range of De between 0 and 10 over the Reynolds number region covering 10^4 to 10^6:

$$\frac{1}{\sqrt{f}} = 3.57 \log \frac{\mathrm{Re}^{\ 1+0.175\mathrm{De}^{(1-0.04\log\mathrm{Re})}}}{(6.5)^{1+0.725\mathrm{De}^{0.7}}} \tag{2.24}$$

Equation (2.22) certainly gives the correct description of the velocity profile during turbulent flow of drag reducing fluids, but there is always an inclination to look for simpler expressions that are good approximations of the exact one at least in a limited region. The one-seventh power law in the case of Newtonian fluids was developed with this view in mind. In the case of drag reducing fluids, such a correlation has also been developed (Shenoy & Mashelkar, 1983; Shenoy & Shintre, 1986).

Table 2.1 Values of α and β

De	α	β
0	0.0790	0.25
1	0.0782	0.262
2	0.0767	0.271
3	0.0741	0.278
4	0.0726	0.285
5	0.0689	0.289
6	0.0655	0.292
7	0.0662	0.301
8	0.0687	0.312
9	0.0732	0.324
10	0.0762	0.334
≥20	0.42	0.55

Shenoy and Shintre (1986). Reprinted with permission from Wiley and Copyright Clearance Center through RightsLink

The Blasius-type correlation developed (Shenoy & Mashelkar, 1983) for drag reducing fluids gives the following form for friction factor:

$$f = \frac{\alpha}{\mathrm{Re}^{\beta}} \quad \text{for} \quad 5 \times 10^3 \leq \mathrm{Re} \leq 10^5 \text{ and } 0 \leq \mathrm{De} \leq 10 \tag{2.25}$$

α and β are both functions of Deborah number, and their values for varying De are given (Shenoy & Shintre, 1986) and shown in Table 2.1.

It has been suggested (Virk, 1966) that the maximum drag reduction that can be achieved in practice can be described by a unique asymptote given by $\alpha = 0.42$ and $\beta = 0.55$ in Eq. (2.25). In Table 2.1, these values correspond to the limit $\mathrm{De} \geq 20$ which has been used to denote the maximum drag reduction asymptote.

As can be seen the value of α is almost constant for $1 \leq \mathrm{De} \leq 10$ but jumps by about 600% for a change of De from 10 to 20. This is due to the fact that around $\mathrm{De} = 20$ and beyond it, the f versus Re curve has a sudden change of slope and hence it is not only the value of α but the value of β that undergoes a sudden change. Detailed explanation of the maximum drag reduction asymptote and its uniqueness is available elsewhere (Virk, 1975) in the exhaustive article and can be referred to for details.

Now, noting that the wall shear stress $\tau_W = (f/2)\rho\overline{U}^2$ by definition, the following expression can be written by algebraic adjustment of Eq. (2.25):

$$\tau_W = \frac{\alpha}{2^{1+\beta}} \rho \overline{U}^{2-\beta} \left(\frac{\mu}{\rho}\right)^{\beta} R^{-\beta} \tag{2.26}$$

Using the definition of $u^* = (\tau_W/\rho)^{1/2}$ and the fact that $\overline{U} = \overline{\psi}\, u_m$, Eq. (2.26) can be written as

$$\frac{u_\mathrm{m}}{u^*} = \left(\frac{2^{1+\beta}}{\alpha\,\overline{\psi}^{2-\beta}}\right)^{\frac{1}{2-\beta}} \left(\frac{R\,u^*\,\rho}{\mu}\right)^{\frac{\beta}{2-\beta}} \tag{2.27}$$

The expression for $\overline{\psi}$ is derivable along the lines done for power-law fluids (Skelland, 1967) using the value of $n = 1$ in the suggested expression. Thus,

$$\overline{\psi} = \frac{(2-\beta)^2}{4-\beta} \tag{2.28}$$

The validity of the above expression is on the simple basis that equations for power-law fluids (Skelland, 1967) are valid for Newtonian fluids as well, and hence, if β is a function of De, then Eq. (2.28) should hold for drag reducing fluids which are Newtonian in viscosity with a mild level of elasticity as exemplified by De.

The key assumption for getting the simplified form of the velocity profile lies in an approximation that Eq. (2.27) is valid at any distance y rather than at R specifically. Thus,

$$u^+ = A\,y^{+B} \tag{2.29}$$

where

$$A = \frac{2^{\frac{1+\beta}{2-\beta}}(4-\beta)}{\alpha^{\frac{1}{2-\beta}}(2-\beta)^2} \tag{2.30}$$

$$B = \frac{\beta}{2-\beta} \tag{2.31}$$

Due to the simplifying assumptions made during the derivation of Eq. (2.27), it does not truly give an accurate description of the velocity profile. It does not predict a zero-velocity gradient at the centerline and has a limited range of validity to Reynolds numbers between 5×10^3 and 10^5. Despite all its limitations, it has extensive use in engineering design especially for estimating the turbulent entrance length, as will be shown later.

Several attempts have been made to experimentally measure the velocity profiles during turbulent flow of drag reducing solutions (Astarita & Nicodemo, 1966; Elata et al., 1966; Ernst, 1966; Patterson & Florez, 1969; Pruitt & Crawford, 1965; Shaver & Merrill, 1959; Smith, Merrill, Mickley, & Virk, 1967; Wells, 1965). Initial velocity profile measurements were made using an impact tube, but erroneous evaluations resulted due to the normal stress contribution of the viscoelasticity of drag reducing fluids leading to lower bulk mean velocities when integrated over the pipe area. The extent of the error involved depended upon the magnitude of the normal stress effect and hence, on the type, concentration, and effective molecular weight of the drag reducing additive. Errors of less than 7% were found (Ernst, 1966; Wells, 1965) as against those which were so large that the apparent velocity head was negative near the wall (Pruitt & Crawford, 1965). Errors as large as 29% were obtained during velocity measurements of flowing drag reducing fluids using the pitot tube (Astarita & Nicodemo, 1966). No calibration procedure for the velocity profile measurements by impact tubes could be used because the normal stress effects of the drag reducing fluid varied with wall shear rate and over the radial profile. To overcome these limitations of the impact tubes, other techniques were tried (Friehe & Schwarz, 1969; Patterson & Florez, 1969; Seyer & Metzner, 1967a; Seyer & Metzner, 1967b; Smith et al., 1967) for velocity profile measurements such as hot-film anemometry and the gas bubble tracer technique.

In hot-film anemometry, the heat transfer rate from the hot film to the fluid primarily depends on the fluid velocity, and hence, a calibration procedure for velocity versus heat transfer rate correlation could be specified. A radial variation in cylindrical hot-film sensitivity was observed (Smith et al., 1967) during measurements which seemed to disappear with degradation of the polymer. However, since their conclusions were based on comparisons with data from the impact tube, their results cannot be taken at face value. Velocity profiles were measured (Patterson & Florez, 1969) during drag reduction flow of viscoelastic polymer and soap solutions using hot-film anemometry. The u^+ — y^+ plot obtained by them for 0.2% polyisobutylene in cyclohexane is shown in Fig. 2.2.

There is an obvious strong tendency toward viscous boundary-layer thickening as observed (Ernst, 1966; Wells, 1965) and predicted (Elata et al., 1966). In more concentrated solutions, such a tendency was found to be stronger. However, soap solution velocity profiles showed (Patterson & Florez, 1969) a different kind of behavior as compared to polymer solutions.

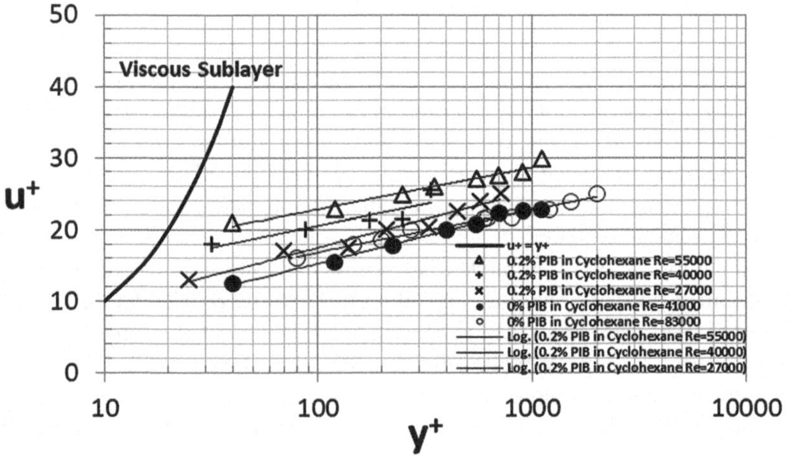

Fig. 2.2 Velocity profiles for 0.2% polyisobutylene in cyclohexane (Patterson & Florez, 1969). (Reprinted with permission from Springer Nature and Copyright Clearance Center through RightsLink)

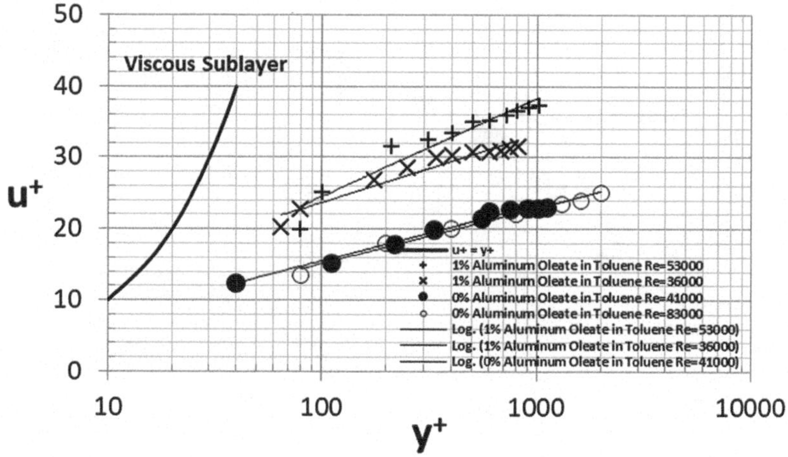

Fig. 2.3 Velocity profiles for 1% aluminum dioleate in toluene (Patterson & Florez, 1969). (Reprinted with permission from Springer Nature and Copyright Clearance Center through RightsLink)

Figure 2.3 shows a plot of u^+ versus y^+ for 1% aluminum dioleate in toluene, clearly indicating no tendency for viscous boundary-layer thickening. The reason for this mainly lies in the fact that the mechanism of drag reduction is different in micellar systems than in polymeric systems.

Velocity profile measurements have also been reported during turbulent flow of drag reducing solutions using a tracer technique (Seyer & Metzner, 1967b). Small gas bubbles were photographed intermittently, and their streak lengths on the film were used as a measure of the velocity. Since the movement of bubbles relative to the fluid was slow, the average velocity measured by averaging many streak lengths at a given point was quite accurate to an extent of less than 3% error.

Figure 2.4 shows the results of their velocity profile measurements, indicating the increasing viscous sublayer thickening with increasing polymer concentration as in Fig. 2.2.

In general, velocity profile measurements are relatively difficult experimentally. Though hot-film anemometers are an improvement over the pitot tubes, it has been shown (Fabula, 1966; Friehe & Schwarz, 1969; James, 1967) that cylindrical hot-film anemometer probes cannot be used to measure velocity under certain conditions. The anomalous effects seem to be associated with stagnation points, as was indicated by results with conical probes which showed that there was no great difference between drag reducing solutions and water (Friehe & Schwarz, 1969).

Developing Flow Entrance Lengths

The velocity profiles discussed in the foregoing section all refer to the fully developed turbulent flow region. However, there is always a region upstream of the flow

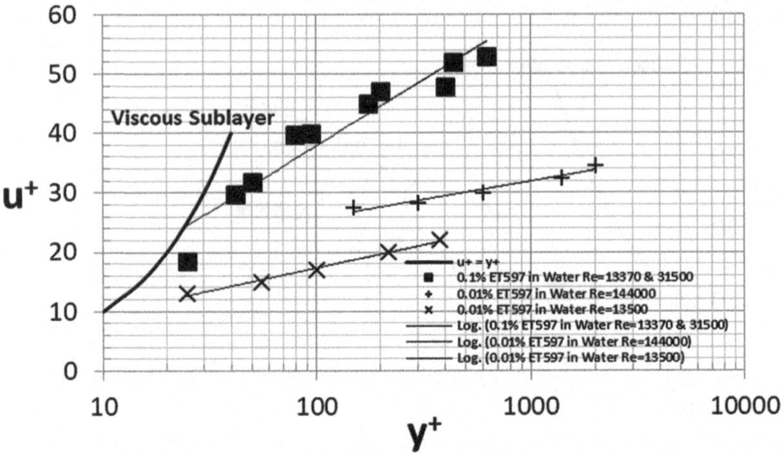

Fig. 2.4 Velocity profiles for ET597 in water (Patterson & Florez, 1969). (Data was taken from Seyer and Metzner (1967b). Reprinted with permission from Springer Nature and Copyright Clearance Center through RightsLink)

where the fluid is in the developing region and becomes fully developed only after a certain entrance length. Since the velocity distributions in the entrance region are significantly different from those in the fully developed region, one would need to have an estimate of the entrance length. For drag reducing fluids, a theoretical prediction along with an experimental verification of the hydrodynamic entrance length during turbulent flow has been done (Shintre, Mashelkar, & Ulbrecht, 1977). A numerical integration technique was used to obtain the variation of entrance with Reynolds number for different Deborah numbers. A simple and elegant approach was used (Shenoy & Mashelkar, 1983) to obtain reasonable estimates of the hydrodynamic entrance lengths for turbulent flow of drag reducing fluids. Close agreement between the results of the two approaches (Shenoy & Mashelkar, 1983; Shintre et al., 1977) was found, and the straightforward method (Shenoy & Mashelkar, 1983) based on an ordering technique is discussed in the following.

The simple form of the velocity profile as given by Eq. (2.27) is assumed to be applicable to the edge of the boundary layer (i.e., $y = \delta$ where $u = u_e$), thus giving

$$\frac{u_e}{u^*} = \frac{2^{\frac{1+\beta}{2-\beta}}(4-\beta)}{\alpha^{\frac{1}{2-\beta}}(2-\beta)^2}\left[\frac{\delta u^* \rho}{\mu}\right]^{\frac{\beta}{2-\beta}} \tag{2.32}$$

Rearranging Eq. (2.32) and using $\tau_w = \rho\, u^{*2}$ give

$$\tau_w = \frac{\alpha}{2^{1+\beta}}\left[\frac{(2-\beta)^2}{(4-\beta)}\right]^{2-\beta} \rho u_e^{2-\beta}\left[\frac{\mu}{\rho\delta}\right]^{\beta} \tag{2.33}$$

An alternative form of τ_w in the turbulent boundary layer for drag reducing fluids can be written as

$$\tau_w = \psi_1 \rho u_e^2 \left[\frac{d\delta}{dx}\right]^{\beta} \tag{2.34}$$

where

$$\psi_1 = \frac{\beta(1+\beta)}{2(2+\beta)} \tag{2.35}$$

Comparing Eqs. (2.33) and (2.34) and solving for δ give

$$\delta = \left\{\frac{(1+\beta)\alpha\overline{\psi}^{2-\beta}}{\psi_1 \ 2^{1+\beta}}\right\}^{\frac{1}{1+\beta}} \left[\frac{\mu}{\rho u_e}\right]^{\frac{\beta}{1+\beta}} \left[\frac{1}{1+\beta}\right] x \tag{2.36}$$

To get an estimate of the turbulent entry length, x is set equal to x_e (the entrance length). δ is set equal to $D/2$ (the pipe centerline), and u_e is set equal to u_m (the maximum axial velocity at pipe centerline $= \overline{U}/\overline{\psi}$). Then, by simplification and substitution of the expressions for $\overline{\psi}$ and ψ_1 from Eqs. (2.28) and (2.35), the following expression for the turbulent entrance length for drag reducing fluids in smooth circular pipes is derived:

$$\frac{x_e}{D} = \left\{\frac{\beta(4-\beta)^2}{2\alpha(2+\beta)(2-\beta)^4}\right\} \mathrm{Re}^{\beta} \tag{2.37}$$

Figure 2.5 shows a plot of the dimensionless turbulent entrance lengths for flow of drag reducing fluids through smooth straight circular pipes. Experimental data on turbulent entrance lengths have been provided (Seyer & Catania, 1972; Shintre et al., 1977). A comparison of the theoretical predictions and the experimental data with approximately estimated De showed satisfactory agreement. At maximum drag reduction asymptote, the entrance lengths are significantly larger, and hence, care must be exercised when measuring velocity profiles of effective drag reducing solutions to make sure that it is in the fully developed regime.

Annular Ducts

In industrial practice, there are likely to be several situations wherein a drag reducing fluid enters an annular space from a large upstream reservoir, resulting in a developing flow which becomes fully developed only after a certain entrance length. Since the velocity and pressure distributions in the entrance region are significantly

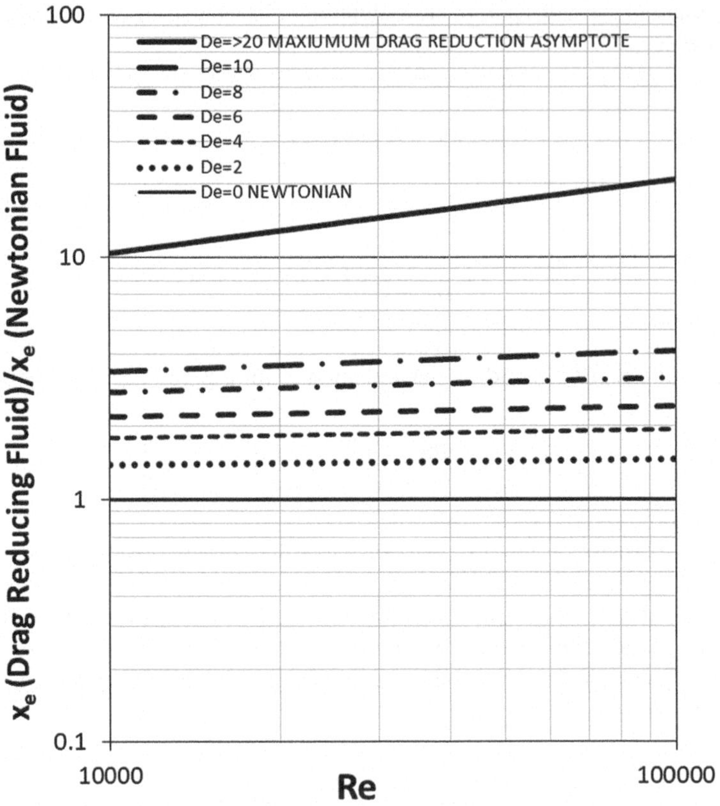

Fig. 2.5 Estimated normalized turbulent entrance length ratios as a function of Reynolds and Deborah numbers for smooth straight circular pipes (Shenoy & Mashelkar, 1983). (Reprinted with permission from American Chemistry Society and Copyright Clearance Center through RightsLink)

different from those in the fully developed flow, one would need to have an estimate of the entrance region in the annuli.

So far as turbulent entrance region flow of non-Newtonian fluids is concerned, the significant available work (Shenoy & Shintre, 1986) deals with developing and fully developed flow turbulent flow of drag reducing fluids in an annular duct. Design equations have been provided for estimating the hydrodynamic turbulent entrance lengths, and expressions are given for fully developed velocity profiles for flow of mildly elastic drag reducing fluids through annular space. The approach is based on a simple ordering approximation. Such a simple analysis of entrance region flow is useful in the absence of an exact solution of the hydrodynamic problem of turbulent flow of drag reducing fluids.

Fully Developed Velocity Profiles

The expression for friction factor for turbulent flow of drag reducing fluids has been given (Seyer & Metzner, 1969b) as shown in Eq. (2.13). By straightforward manipulations, an explicit *f*(Re, De) function is obtained as shown in Eq. (2.25). Noting that the wall shear stress $= (f/2)\rho\overline{U}^2$ by definition, Eq. (2.25) is written by algebraic adjustment as Eq. (2.26). Using the definition of $u^* = (\tau_w/\rho)^{1/2}$ and the fact that $\overline{U} = \overline{\psi} u_m$, Eq. (2.26) is rewritten as Eq. (2.27). The expression for ψ is derivable along the lines for power-law fluids (Skelland, 1967) using the value of $n = 1$ in the suggested expression and given as Eq. (2.28). The validity of Eq. (2.28) is simply on the basis that Skelland's equations for power-law fluids are valid for Newtonian fluids as well. Since drag reducing fluids are assumed to be Newtonian in viscosity but have mild elasticity, the equations in the Newtonian form would hold except for altered values of α and β which would now have to be functions of elasticity rather than of pseudoplasticity. An approximation is now made that Eq. (2.27) is valid for any wall distance y and, therefore, can be written as given in Eq. (2.29).

The values of *A* and *B* for various values of α and β corresponding to different Deborah numbers are given in Table 2.2.

Note that Eq. (2.29) is applicable for flow through a smooth circular straight pipe and as such is not applicable to the annulus.

It is now assumed that the effect of the inner cylinder on the velocity profile in the outer region is small enough to be neglected. Thus, the expression for the velocity profile given by Eq. (2.29) is held valid for the outer region of the annulus (Fig. 2.6) and written as

Table 2.2 Values of α and β as well as *A* and *B*

De	α	β	*A*	*B*
0	0.0790	0.25	8.568	0.143
1	0.0782	0.262	8.870	0.151
2	0.0767	0.271	9.168	0.157
3	0.0741	0.278	9.516	0.161
4	0.0726	0.285	9.799	0.166
5	0.0689	0.289	10.20	0.169
6	0.0655	0.292	10.59	0.171
7	0.0662	0.301	10.77	0.177
8	0.0687	0.312	10.84	0.185
9	0.0732	0.324	10.77	0.193
10	0.0762	0.334	10.79	0.200
≥20	0.42	0.55	6.26	0.379

Shenoy and Shintre (1986). Reprinted with permission from Wiley and Copyright Clearance Center through RightsLink

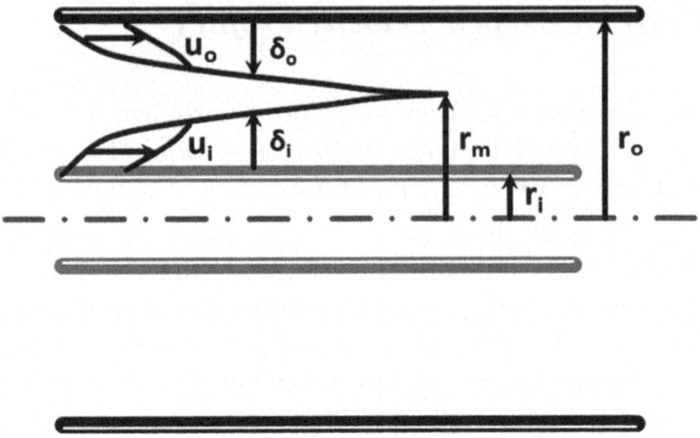

Fig. 2.6 Schematic diagram of the annular entrance region flow (Shenoy & Shintre, 1986). (Reprinted with permission from Wiley and Copyright Clearance Center through RightsLink)

$$u_o^+ = Ay_o^{+B} \tag{2.38}$$

where the subscript o represents the outer region of the annulus. An expression of the similar form is assumed to hold for the inner region, thus giving

$$u_i^+ = A_i y_i^{+B_i} \tag{2.39}$$

To obtain expressions for A_i and B_i, tedious algebra would have to be followed leading to expressions which would be too cumbersome for engineering applications. The situation, however, becomes greatly simplified if it is assumed that the slopes of the u_o^+ versus y_o^+ and the u_i^+ versus y_i^+ are the same, thus giving $B_i = B$.

The momentum balance for the inner and outer flow region of the annuli gives the following expressions for shear stresses τ_{wi} and τ_{wo}:

$$\tau_{wi} = \frac{r_i}{2}\left[\frac{\lambda^2}{\kappa^2} - 1\right]\frac{\Delta P}{L} \tag{2.40}$$

$$\tau_{wo} = \frac{r_o}{2}\left[1 - \lambda^2\right]\frac{\Delta P}{L} \tag{2.41}$$

where

$$\lambda = \frac{r_m}{r_o}; \qquad \kappa = \frac{r_i}{r_o} \tag{2.42}$$

Using the definition of $u^* = (\tau_w/\rho)^{½}$ for the inner and outer region, the following can be easily derived. Thus,

$$\frac{u_o^*}{u_i^*} = \left[\frac{\kappa\left(1 - \lambda^2\right)}{\left(\lambda^2 - \kappa^2\right)}\right]^{1/2} \tag{2.43}$$

The expression for A_i in Eq. (2.39) can be derived, using the fact that $u_i = u_o$ at $r = r_m$ where r_m is the radius of maximum velocity. Dividing Eq. (2.38) by Eq. (2.39) and using Eq. (2.43), the expression for A_i can be obtained as

$$A_i = A\left[\frac{\kappa(1-\lambda^2)}{(\lambda^2-\kappa^2)}\right]^{\frac{B+1}{2}}\left[\frac{(1-\lambda)}{(\lambda-\kappa)}\right]^{B} \tag{2.44}$$

Thus, the velocity profile for the inner region of the annulus can be written in the following form:

$$u_i^+ = A\left[\frac{\kappa(1-\lambda^2)}{(\lambda^2-\kappa^2)}\right]^{\frac{B+1}{2}}\left[\frac{(1-\lambda)}{(\lambda-\kappa)}\right]^{B}\left(y_i^+\right)^B \tag{2.45}$$

For Newtonian fluids, Eqs. (2.38) and (2.45) reduce to

$$u_o^+ = 8.57{y_o^+}^{1/7} \tag{2.46}$$

and

$$u_i^+ = 8.57\left[\frac{\kappa(1-\lambda^2)}{(\lambda^2-\kappa^2)}\right]^{4/7}\left[\frac{(1-\lambda)}{(\lambda-\kappa)}\right]^{1/7}\left(y_i^+\right)^{1/7} \tag{2.47}$$

These expressions are akin to the expressions provided (Singh, Nigam, & Mishra, 1980) for the Newtonian turbulent velocity profiles in the outer and inner regions of an annulus, except for the value of A which is slightly different, and explanation for such deviation is available (Schlichting, 1960). Equations (2.38) and (2.45) define the complete velocity profile for drag reducing fluids in an annulus.

Friction Factors

Since the shear stresses at the inner and outer surfaces in an annulus are unequal as can be seen from Eqs. (2.40) and (2.41) and their variation is nonlinear in the radial direction, the friction factors in the two regions would be different. Having established the fully developed velocity profiles, the relationships between the friction factor and Reynolds number in the corresponding flow regions can be easily derived as follows:

$$f_i = \frac{2\tau_{wi}}{\rho\overline{U}_i^2} \text{ and } \mathrm{Re}_i = \frac{D_{ei}\overline{U}_i}{\upsilon} \tag{2.48}$$

$$f_{\mathrm{o}} = \frac{2\tau_{\mathrm{wo}}}{\rho \overline{U}_{\mathrm{o}}^{2}} \text{ and } \mathrm{Re}_{\mathrm{o}} = \frac{D_{\mathrm{eo}} \overline{U}_{\mathrm{o}}}{\upsilon} \tag{2.49}$$

where

$$D_{\mathrm{ei}} = 2r_{\mathrm{i}} \left[\frac{\lambda^2}{\kappa^2} - 1 \right] \tag{2.50}$$

and

$$D_{\mathrm{eo}} = 2r_{\mathrm{o}} \left[1 - \lambda^2 \right] \tag{2.51}$$

The expressions for the average velocity $\overline{U}$ for the inner and outer region can be obtained from the following definition using the appropriate expressions for the velocity profiles in the inner and outer region of the annulus as given by Eqs. (2.38) and (2.45):

$$\overline{U}_{\mathrm{o}} = \frac{\int_{r_{\mathrm{m}}}^{r_{\mathrm{o}}} u_{\mathrm{o}} 2\pi r \, dr}{\int_{r_{\mathrm{m}}}^{r_{\mathrm{o}}} 2\pi r \, dr}; \quad \overline{U}_{\mathrm{i}} = \frac{\int_{r_{\mathrm{i}}}^{r_{\mathrm{m}}} u_{\mathrm{i}} 2\pi r \, dr}{\int_{r_{\mathrm{i}}}^{r_{\mathrm{m}}} 2\pi r \, dr} \tag{2.52}$$

Then, using the definition that

$$\frac{u^*}{\overline{U}} = \sqrt{\frac{f}{2}} \tag{2.53}$$

expressions for f_{i} and f_{o} can be derived as

$$f_{\mathrm{i}} = \frac{\Phi_i}{\mathrm{Re}_i^{\beta}}; \quad f_{\mathrm{o}} = \frac{\Phi_{\mathrm{o}}}{\mathrm{Re}_{\mathrm{o}}^{\beta}} \tag{2.54}$$

where

$$\Phi_{\mathrm{i}} = \left\{ \frac{(B+1)\, 2^{\frac{3B-1}{2}}}{A} \left[\frac{\lambda^2 - \kappa^2}{\kappa(1-\lambda^2)} \right]^{\frac{B+1}{2}} \times \left[\frac{\lambda - \kappa}{\kappa(1-\lambda)} \right]^{B} \frac{(\lambda+\kappa)^{B+1}}{\left[\lambda - \frac{\lambda-\kappa}{B+2} \right]} \right\}^{\frac{2}{B+1}} \tag{2.55}$$

and

$$\Phi_{\mathrm{o}} = \left\{ \frac{(B+1)\, 2^{\frac{3B-1}{2}}}{A} \frac{(1+\lambda)^{B+1}}{\left[\lambda + \frac{1-\lambda}{B+2} \right]} \right\}^{\frac{2}{B+1}} \tag{2.56}$$

For Newtonian fluids

$$f_{\rm i} = \frac{\Phi_{\rm i}}{{\rm Re}_{\rm i}^{0.25}}; \quad f_{\rm o} = \frac{\Phi_{\rm o}}{{\rm Re}_{\rm o}^{0.25}} \tag{2.57}$$

where

$$\Phi_{\rm i} = 0.0208\left[\frac{\lambda^2 - \kappa^2}{\kappa(1-\lambda^2)}\right] \times \left[\frac{\lambda - \kappa}{\kappa(1-\lambda)}\right]^{0.25} \frac{(\lambda + \kappa)^2}{\left[\lambda - \frac{7}{15}(1-\lambda)\right]^{1.75}} \tag{2.58}$$

and

$$\Phi_{\rm o} = 0.0208\frac{(1+\lambda)^2}{\left[\lambda + \frac{7}{15}(1-\lambda)\right]^{1.75}} \tag{2.59}$$

Equations (2.58) and (2.59) are the same as those derived for Newtonian fluids (Singh et al., 1980) except for the coefficient which is different in the fourth-decimal place. This is due to the value of A being taken as 8.57 in the above as against 8.74 (Singh et al., 1980).

Developing Flow Entrance Lengths

As noted earlier, the shear stresses at the inner and outer wall are different, and in fact, the stress at the inner wall is greater than that at the outer wall. The development of the velocity profiles in the two corresponding regions are thus bound to be different. Hence, two boundary-layer thicknesses developing simultaneously could be expected, one starting from the inner wall, namely, $\delta_{\rm i}$, and the other at the outer wall, namely, $\delta_{\rm o}$.

It is assumed that the velocity profiles given by Eqs. (2.38) and (2.45) are valid at the edge of the corresponding boundary layers. This assumption may not be strictly valid but, nevertheless, would provide a useful step in the estimation of the entrance lengths. Thus,

$$\frac{u_{\rm o}}{u_{\rm o}^*} = A\left(\frac{\delta_{\rm o} u_{\rm o}^*}{\upsilon}\right)^B \tag{2.60}$$

and

$$\frac{u_{\rm i}}{u_{\rm i}^*} = A_{\rm i}\left(\frac{\delta_{\rm i} u_{\rm i}^*}{\upsilon}\right)^B \tag{2.61}$$

Rearranging Eqs. (2.60) and (2.61), and using the fact that $\tau_w = \rho u^{*2}$, results in the following expressions for the shear stresses in the two regions:

$$\tau_{wi} = \rho \left[\frac{u_i}{A_i}\right]^{\frac{2}{B+1}} \left[\frac{\upsilon}{\delta_i}\right]^{\frac{2B}{B+1}} \tag{2.62}$$

$$\tau_{wo} = \rho \left[\frac{u_o}{A}\right]^{\frac{2}{B+1}} \left[\frac{\upsilon}{\delta_o}\right]^{\frac{2B}{B+1}} \tag{2.63}$$

An alternate form for τ_w has been provided (Skelland, 1967) for turbulent boundary layer, and this can be written for the inner and outer regions as follows:

$$\tau_{wi} = \psi_1 \rho u_i^2 \left(\frac{d\delta_i}{dx}\right) \tag{2.64}$$

$$\tau_{wo} = \psi_1 \rho u_o^2 \left(\frac{d\delta_o}{dx}\right) \tag{2.65}$$

where

$$\psi_1 = \frac{\beta(2-\beta)}{2(2+\beta)} \tag{2.66}$$

Comparing Eqs. (2.62) and (2.63) with Eqs. (2.64) and (2.65) expressions for δ_i and δ_o can be obtained by simple integration to give

$$\delta_i = \left[\frac{x_i(3B+1)}{\psi_1(B+1)}\right]^{\frac{(B+1)}{(3B+1)}} \left[\frac{1}{A_i}\right]^{\frac{2}{(3B+1)}} \left[\frac{\upsilon}{u_i}\right]^{\frac{2B}{(3B+1)}} \tag{2.67}$$

$$\delta_o = \left[\frac{x_o(3B+1)}{\psi_1(B+1)}\right]^{\frac{(B+1)}{(3B+1)}} \left[\frac{1}{A}\right]^{\frac{2}{(3B+1)}} \left[\frac{\upsilon}{u_o}\right]^{\frac{2B}{(3B+1)}} \tag{2.68}$$

To get an estimate of the turbulent entry length, x_i is set to be equal to x_{ei} at $\delta_i = r_m - r_i$ and $u_i = u_m$ (maximum velocity at $r = r_m$), and x_o is set to be equal to x_{eo} at $\delta_o = r_o - r_m$ and $u_o = u_m$. The maximum velocity is proportional to the average velocity (Skelland, 1967) which for a smooth pipe is equal to $\overline{U}/\overline{\psi}$ where $\overline{\psi}$ is given by Eq. (2.28).

The expressions for the entrance length after considerable simplifications and substitution of the expressions for A, B, A_i, $\overline{\psi}$, and ψ_1 from Eqs. (2.30), (2.31), (2.44), (2.28), and (2.35) yield

$$\frac{x_{ei}}{F_i D_{ei}} = \left\{ \frac{\beta(4-\beta)^2}{2\alpha(2+\beta)(2-\beta)^4} \right\} \mathrm{Re}_i^{\beta} \tag{2.69}$$

$$\frac{x_{eo}}{F_o D_{eo}} = \left\{ \frac{\beta(4-\beta)^2}{2\alpha(2+\beta)(2-\beta)^4} \right\} \mathrm{Re}_o^{\beta} \tag{2.70}$$

where

$$\mathrm{Re}_i = \frac{D_{ei} U}{\upsilon}; \quad \mathrm{Re}_o = \frac{D_{eo} U}{\upsilon} \tag{2.71}$$

$$F_i = \left[\frac{(1-\lambda^2)}{(\lambda+\kappa)} \right] \times \left[\frac{\kappa(1-\lambda)}{(\lambda^2-\kappa^2)} \right]^{\beta} \tag{2.72}$$

$$F_o = \frac{(1-\lambda)}{(1+\lambda)^{\beta}} \tag{2.73}$$

Due to the difference in the velocity profiles in the two flow regimes, the entrance lengths for the inner and outer regions are different. In fact, if the expressions for F_i, F_o, D_{ei}, and D_{eo} are substituted and proper simplification is carried out, then it is seen that X_{ei} and X_{eo} are related as follows:

$$X_{ei} = \left[\frac{(\lambda-\kappa)}{\kappa(1-\lambda)} \right] X_{eo} \tag{2.74}$$

Thus, $X_{ei} > X_{eo}$ as the multiplying factor above is a whole number greater than one. Thus, from design considerations, it is advantageous to base all calculations on X_{ei}, as it guarantees fully developed flow beyond this value.

The radius of maximum velocity is determined (Singh et al., 1980) as

$$\left[\frac{(\lambda-\kappa)}{(1-\lambda)} \right] = \kappa^{0.343} \tag{2.75}$$

This equation is quite general as can be seen by its application to the two limiting cases of the annulus geometry, namely, a circular pipe and a parallel plate channel. In the case of a circular pipe, $\kappa = 0$ and $\lambda = 0$ irrespective of the type of fluid. Equation (2.75) does indeed give $\lambda = 0$ for a circular pipe and in the case of a parallel plate channel, $\lambda = 1$ irrespective of the type of fluid.

Figure 2.7 shows a plot of $X_{ei}/F_i D_i$, $X_{eo}/F_o D_o$ as functions of Reynolds number ($10^4 \leq \mathrm{Re} \leq 10^5$) for varying Deborah number. It is seen that the entrance lengths are increasing with increasing Deborah as well as Reynolds numbers. In fact, for conditions of maximum drag reduction, the entrance lengths are an order of magnitude higher than that for the Newtonian case. The maximum drag reduction

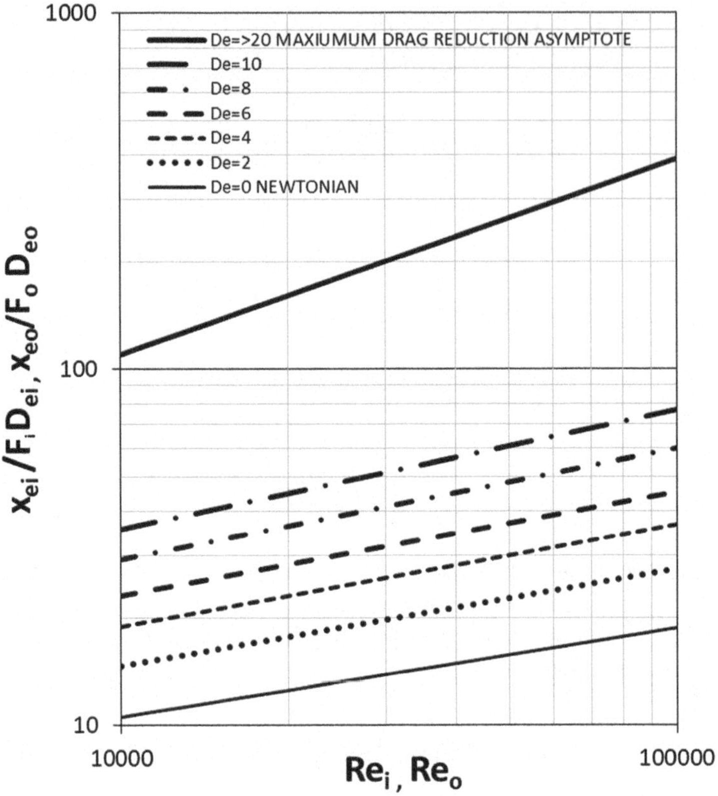

Fig. 2.7 Entrance lengths as functions of Reynolds and Deborah numbers (Shenoy & Shintre, 1986). (Reprinted with permission from Wiley and Copyright Clearance Center through RightsLink)

corresponds to a Deborah number 20, thus giving an estimate of the level of entrance length difference that elasticity alone can create.

It would have been beneficial if the theoretical predictions were compared with reliable experimental data. While numerical simulation of drag reducing turbulent flow in annular conduits has been done (Azouz & Shirazi, 1997), the other available information in the literature on turbulent flow of non-Newtonian fluid through an annulus (Tiu, 1979) contains no data on the entrance lengths or velocity profiles. In such cases, the only option is to check the validity of the theoretical findings in the limiting case. The simplest limiting case of annular flow is a cylindrical pipe obtained by taking $F_o = l$, for which the theoretical predictions are akin to those available (Shintre et al., 1977), and hence an experimental verification would only be duplication of information, as this has already been done (Shintre et al., 1977) for drag reducing fluids.

Experimental values of tube flow friction factors are available for several polymer solutions (Darby & Chang, 1984). Correlations to calculate the Deborah numbers

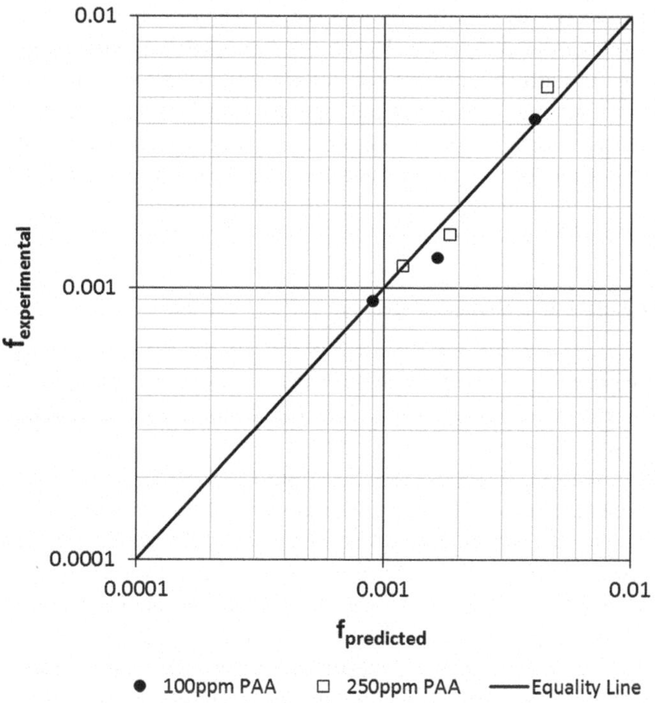

Fig. 2.8 Comparison of the theoretical predictions of computed friction factors with the tube flow experimental data. [●100 ppm PAA; □250 ppm PAA] of Darby and Chang (1984). (Shenoy & Shintre, 1986). (Reprinted with permission from Wiley and Copyright Clearance Center through RightsLink)

and the relaxation times for these solutions are also available (Darby & Chang, 1984). Using these correlations and the rheological data, the Deborah numbers can be estimated. From Table 2.1, values of α and β corresponding to the estimated Deborah numbers are got, and friction factor values are computed. Figure 2.8 shows a plot of computed friction factors (Shenoy & Shintre, 1986) versus the experimental friction factors (Darby & Chang, 1984), and the agreement is seen to be very good. The following data (Darby & Chang, 1984) was used to establish this verification of the results of theoretical prediction (Shenoy & Shintre, 1986) with experimental results:

Polymer solutions: 100 ppm and 250 ppm Separan AP-30
Tube diameter: 0.406 cm inside diameter
Solvent Reynolds numbers: 10^4, 5×10^4 and 10^5
Solvent: water

Though the theoretical predictions (Shenoy & Shintre, 1986) have not been validated for annular flow, the order of magnitude should provide results adequate enough for design purposes. Experimental verification is, undoubtedly, difficult because of the uncertainty in estimation of correct value of Deborah number through reliable fluid relaxation time data. However, the results given in Fig. 2.7 for the maximum drag reduction asymptote can certainly be used for a conservative design. Since the maximum drag reduction asymptote is itself independent of the system variables such as dimensions of the annulus, polymer type, and concentration (Virk, 1975), the predictions based on it would carry the same advantages.

Rough Straight Circular Pipes

The velocity profiles developed in the foregoing sections had an underlying assumption that the pipes considered were free from defects on the inside and hence could be considered as smooth.

However, in reality, all pipes have a certain level of surface defects marked by protrusions or indentations. The actual dimensions of this roughness do not have as great an effect as the relative roughness due to the size of the protrusions or indentations in comparison with the dimensions of the pipe. If the average height of the roughness projections is expressed as ε, then the relative roughness factor is written as ε/D, and the relative roughness factor for smooth pipes is zero.

A rough surface does not always act rough but depends on the relative magnitude of the size of the surface elements and the thickness of the viscous sublayer when the fluid is flowing under turbulent conditions. As the Reynolds number is increased, the relative thickness of the viscous sublayer decreases, and eventually the roughness elements protrude through the sublayer. When this occurs, the predominant source of pressure drop becomes form drag caused by the roughness elements, and the flow is then termed "fully rough."

Fully Developed Velocity Profiles

The velocity profile in rough pipes is normally obtained by simply adding terms to the Newtonian law of the wall (Fenter, 1959; Spangler, 1969). The latter provided the expression for drag reducing fluids by modifying the suggested profile (Meyer, 1966) in Eq. (2.5). Thus,

$$u^+ = 2.5 \ln y^+ + 5.5 + \alpha^* \ln \left(\frac{u^*}{u^*_{\text{crit}}} \right) - F \tag{2.76}$$

where the roughness function F is defined as

$$F = \sqrt{2}\left\{\left(1.77 + \frac{\alpha^*}{\sqrt{2}}\right) \ln \; \mathrm{Re}_{\mathrm{w}}\sqrt{f} - \frac{1}{\sqrt{f}} - 0.394 - \frac{\alpha^*}{\sqrt{2}} \ln \frac{\sqrt{2}D\rho u^*_{\mathrm{crit}}}{\mu}\right\} \quad (2.77)$$

Equation (2.77) allows the determination of the value of the roughness function for a drag reducing fluid from friction factor – Reynolds number data taken in a rough pipe when α^* and u^*_{crit} for the fluid are known. The roughness functions as experimentally obtained (Spangler, 1969) for water and polymer drag reducing fluid (31 wppm P-295 in water) are shown in Figs. 2.9 and 2.10.

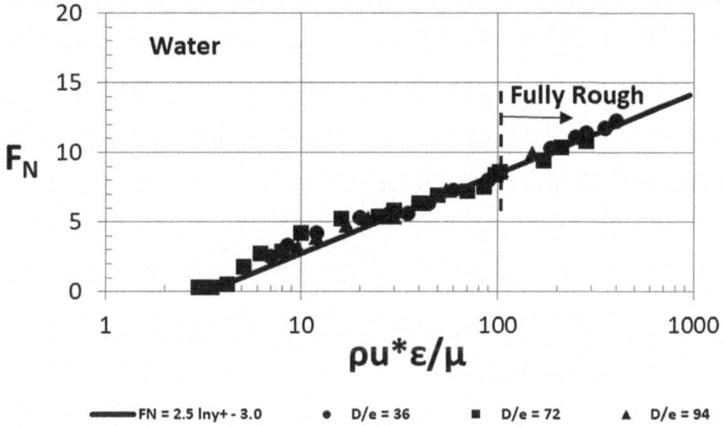

Fig. 2.9 Roughness function for water (Spangler, 1969). (Reprinted with permission from Springer Nature and Copyright Clearance Center through RightsLink)

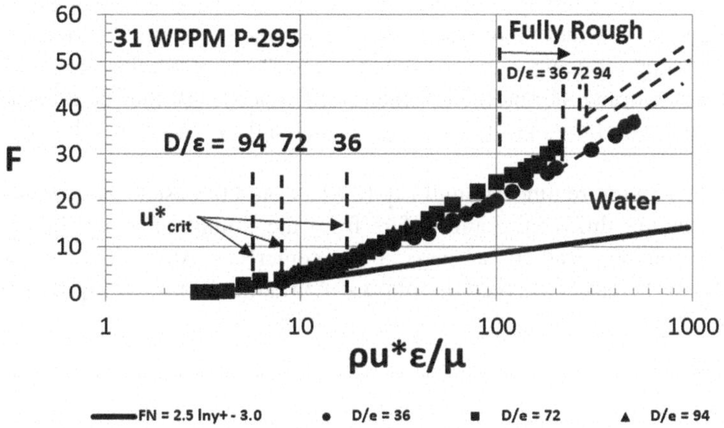

Fig. 2.10 Roughness function for drag reducing fluid (Spangler, 1969). (Reprinted with permission from Springer Nature and Copyright Clearance Center through RightsLink)

It can be seen that the roughness function F for drag reducing fluids becomes larger than the roughness function F_N for Newtonian fluids after a critical shear stress is exceeded. Moreover, for drag reducing fluids, F is found to depend on the value of D/ε, whereas for Newtonian fluids, it does not. Another approach for obtaining F based on the experimental findings is to assume the following form for Eq. (2.77):

$$F = F_N + \alpha^* \ln \frac{u^*}{u^*_{crit}} \quad \text{for } u^* > u^*_{crit} \tag{2.78}$$

For Newtonian fluids, it is clear (Fenter, 1959) that the expression for F_N in the fully rough regime can be given exactly by

$$F_N = 2.5 \ln \frac{\rho u^* \varepsilon}{\mu} - 3.0 \tag{2.79}$$

Combining Eqs. (2.78) and (2.79) gives

$$F = (2.5 + \alpha^*) \ln \frac{\rho u^* \varepsilon}{\mu} - \alpha^* \ln \frac{\rho u^*_{crit} \epsilon}{\mu} \tag{2.80}$$

Combining Eqs. (2.76) and (2.80) gives the conventional Newtonian form for the velocity profile in the rough pipes as

$$u^+ = 2.5 \ln \frac{y^+}{\varepsilon} + 8.5 \tag{2.81}$$

This is because in the fully rough regime all the drag reduction effectiveness of the additive is annulled by the pipe roughness, and hence, the velocity profile can be expected to be no different from the Newtonian case. In very rough pipes, it was found (White, 1967) that there was insignificant change in the frictional characteristics with drag reducing agents even at very large Reynolds numbers. Same conclusion was drawn (Brandt, McDonald, & Boyle, 1969) that drag reducers had little or no effect on flow in fully rough regimes, but they tended to delay the onset of this regime.

In the fully rough regime, though Eq. (2.81) would represent the velocity profile for drag reducing fluids, it again suffers from the limitation of not being able to predict the velocity gradient as zero at the centerline. An expression similar to Eq. (2.22) can be easily derived as follows. A modified form of Eq. (2.22) can be written as

$$u^{+} = 2.46\left[\ln\frac{y}{\varepsilon} + \ln\frac{\rho u^{*}\varepsilon}{\mu} + \left(0.4398 + 0.123\ \mathrm{De} + 0.0135\ \mathrm{De}^{2}\right)\right.$$
$$\left.\times\ \exp\left\{-\frac{(\xi - 0.8)^{2}}{0.129(1 + 0.2\mathrm{De})^{2}}\right\} + 1.3676\left(1 - 0.09\ \mathrm{De} - 0.01\ \mathrm{De}^{2}\right)\right]$$
$$+ 5.6 + 1.55\mathrm{De} - G \tag{2.82}$$

The term 2.46 $\ln\frac{\rho u^{*}\varepsilon}{\mu}$ assumes a value of 3.0, as obtained by comparing Eq. (2.81) with the following expression for Newtonian fluids in smooth pipes:

$$u^{+} = 2.5\ \ln y^{+} + 5.5 \tag{2.83}$$

Thus, the velocity profile for drag reducing fluids in rough pipes can be written as

$$u^{+} = 2.46\left[\ln\frac{y}{\varepsilon} + \left(0.4398 + 0.123\ \mathrm{De} + 0.0135\ \mathrm{De}^{2}\right)\right.$$
$$\left.\times\ \exp\left\{-\frac{(\xi - 0.8)^{2}}{0.129(1 + 0.2\mathrm{De})^{2}}\right\} + 1.3676\left(1 - 0.09\ \mathrm{De} - 0.01\ \mathrm{De}^{2}\right)\right]$$
$$+ 8.6 + 1.55\mathrm{De} - G \tag{2.84}$$

As the flow approaches the fully rough regime, the drag reduction effectiveness decreases and De → 0. Thus, the expression in the fully rough regime would be as follows:

$$u^{+} = 2.46\ \ln\frac{y}{\varepsilon} + 1.082\ \exp\left\{-\frac{(\xi - 0.8)^{2}}{0.129}\right\} + 7.964 \tag{2.85}$$

References

Astarita, G. (1965). Possible interpretation of the mechanism of drag reduction in viscoelastic liquids. *Industrial & Engineering Chemistry Fundamentals, 4*(3), 354–356.

Astarita, G., & Nicodemo, L. (1966). Velocity distributions and Normal stresses in viscoelastic turbulent pipe flow. *AICHE Journal, 12*, 478–484.

Astarita, G., Greco, G. J., & Nicodemo, L. (1969). A phenomenological interpretation and correlation of drag reduction. *AICHE Journal, 15*, 564–567.

Azouz, I., & Shirazi, S. A. (1997). Numerical simulation of drag reducing turbulent flow in annular conduits. *Transfer ASME Journal Fluids Engineering, 119*(4), 838–846.

Bogue, D. C., & Metzner, A. B. (1963). Velocity profiles in turbulent pipe flow. *Industrial & Engineering Chemistry Fundamentals, 2*, 143–152.

Brandt, H., McDonald, A. T., & Boyle, F. W. (1969). Turbulent skin friction of dilute polymer solutions in rough pipes. In C. S. Wells (Ed.), *Viscous drag reduction* (pp. 159–171). New York: Plenum Press.

Clapp, R. M. (1961). *Intern. developments in heat transfer* (p. 652-61, D-159, D-211-5). New York: ASME.

Darby, R., & Chang, H. D. (1984). Generalized correlation for friction loss in drag reducing polymer solutions. *AICHE Journal, 30*(2), 274–280.

Dodge, D. W., & Metzner, A. B. (1959). Turbulent flow of non-newtonian systems. *AICHE Journal, 5*(2), 189–204.

Elata, C., Lehrer, J., & Kahanovitz, A. (1966). Turbulent shear flow of polymer solutions. *Israel Journal of Technology, 4*(1), 87–95.

Ernst, W. D. (1966). Investigation of turbulent shear flow of dilute aqueous CMC solutions. *AICHE Journal, 12*(3), 581–586.

Fabula, A. G. (1966). *An experimental study of grid turbulence in dilute high-polymer solutions* (PhD Thesis). Pennsylvania State University.

Fenter, F. W. (1959). *The turbulent boundary layer on uniformly rough surfaces at supersonic speeds* (Report No. RE-E9R-2). Vought Research Center, Chance Vought Aircraft Inc.

Friehe, C. A., & Schwarz, W. H. (1969). The use of pitot-static tubes and hot-film anemometers in dilute polymer solutions. In C. S. Wells (Ed.), *Viscous drag reduction* (pp. 281–296). New York: Plenum Press.

Hinze, J. D. (1955). *Turbulence*. New York: McGraw-Hill.

James, D. F. (1967). *Laminar flow of dilute polymer solutions around circular cylinders* (PhD Thesis). California Institute of Technology.

Kilbane, J. K., & Greenkorn, R. A. (1966). *Correlation of friction factors for viscoelastic fluids in tubes, Soc. petrol. engrs*. AIME Paper No. SPE 1679.

Krope, A., Krope, J., & Lipus, L. C. (2005). A model for velocity profile in turbulent boundary layer with drag reducing surfactants. *Applied Rheology, 15*(3), 152–159.

Langhaar, H. L. (1951). *Dimensional analysis and theory of models*. New York: Wiley.

Meter, D. M. (1964). Tube flow of non-newtonian polymer solutions: Part II – turbulent flow. *AICHE Journal, 10*(6), 881–884.

Meyer, W. A. (1966). A correlation of the frictional characteristics for turbulent flow of dilute viscoelastic non-newtonian fluids in pipes. *AICHE Journal, 12*(3), 522–525.

Millikan, C. (1939). A critical discussion of turbulent flows in channels and circular tubes. In *Proc. 5th intern. congr. appl. mech*. New York: Wiley.

Nikuradse, J. (1932). Laws of turbulent flow in smooth pipes (English translation). *NASA, TT F-10*, 359.

Patterson, G. K., & Florez, G. L. (1969). Velocity profiles during drag reduction. In C. S. Wells (Ed.), *Viscous drag reduction* (pp. 233–250). New York: Plenum Press.

Pruitt, G. T., & Crawford, H. R. (1965). *Investigations for the use of additives for the reduction of pressure losses*. Western Company, Contract No. DA-23-072-AMC-209 (T) Final Report.

Ramadan, A., Saasen, A., & Skalle, P. (2004). Application of the minimum transport velocity model for drag-reducing polymers. *The Journal of Petroleum Science and Engineering, 44*(3), 303–316.

Schlichting, H. (1960). *Boundary layer theory*. New York: McGraw Hill.

Sellin, R. H., Hoyt, J. W., & Scrivener, O. (1982). The effect of drag reducing additives on fluid flows and their industrial applications. Part 1: Basic aspects. *Journal of Hydraulic Research, 20* (1), 29–68.

Seyer, F. A., & Catania, P. J. (1972). Laminar and turbulent entry flow of polymer solutions. *The Canadian Journal of Chemical Engineering, 50*(1), 31–36.

Seyer, F. A., & Metzner, A. B. (1967a). Turbulent flow properties of viscoelastic fluids. *The Canadian Journal of Chemical Engineering, 45*(3), 121–126.

Seyer, F. A., & Metzner, A. B. (1967b). *Turbulence phenomena in drag reducing systems, 60th annual AICHE meeting*, New York.

Seyer, F. A., & Metzner, A. B. (1969a). Turbulence phenomena in drag reducing systems. *AICHE Journal, 15*(3), 426–434.

Seyer, F. A., & Metzner, A. B. (1969b). Drag reduction in large tubes and the behavior of annular films of drag reducing fluids. *The Canadian Journal of Chemical Engineering, 47*(6), 525–529.

Shaver, R. G., & Merrill, E. W. (1959). Turbulent flow of pseudoplastic polymer solutions in straight cylindrical tubes. *AICHE Journal, 5*(2), 181–188.

Shenoy, A. V. (1988). Encyclopedia of fluid mechanics, vol. 7, ch. 16. In N. P. Cheremisinoff (Ed.), *Turbulent flow velocity profiles in drag-reducing fluids* (pp. 479–503). Houston, TX: Gulf Publishing Co.

Shenoy, A. V., & Mashelkar, R. A. (1983). Engineering estimate of hydrodynamic entrance lengths in non-newtonian turbulent flow. *Industrial and Engineering Chemistry Process Design and Development, 22*(1), 165–168.

Shenoy, A. V., & Saini, D. R. (1982). A new velocity profile model for turbulent pipe flow of power-law fluids. *The Canadian Journal of Chemical Engineering, 60*(5), 694–696.

Shenoy, A. V., & Shintre, S. N. (1986). Developing and fully developed turbulent flow of drag reducing fluids in an annular duct. *The Canadian Journal of Chemical Engineering, 64*(2), 190–195.

Shenoy, A. V., & Talathi, M. M. (1985). Turbulent pipe flow velocity profile model for drag-reducing fluids. *AICHE Journal, 31*(3), 520–522.

Shintre, S. N., Mashelkar, R. A., & Ulbrecht, J. (1977). An approximate theoretical analysis and experimental verification of turbulent entrance region flow of drag reducing fluids. *Rheologica Acta, 16*(5), 490–496.

Singh, R. P., Nigam, K. K., & Mishra, P. (1980). Developing and fully developed turbulent flow in an annular duct. *Journal of Chemical Engineering of Japan, 13*(5), 349–353.

Skelland, A. H. (1967). *Non-newtonian flow and heat transfer*. New York: Wiley.

Smith, K. A., Merrill, E. W., Mickley, H., & Virk, P. (1967). Anomalous pitot tube and hot film measurements in dilute polymer solutions. *Chemical Engineering Science, 22*(4), 619–626.

Spangler, J. G. (1969). Studies of viscous drag reduction with polymers including turbulence measurements and roughness effects. In C. S. Wells (Ed.), *Viscous drag reduction* (pp. 131–157). New York: Plenum Press.

Stein, M. A., Kessler, D. P., & Greenkorn, R. A. (1980). An empirical model for velocity profiles for turbulent flow in smooth pipes. *AICHE Journal, 26*(2), 308–310.

Tiu, C. (1979). *Turbulent flow behaviour of dilute polymer solutions in an annulus, Soc. Rheol. Golden Jubilee Meeting.*

Virk, P. S. (1966). *The toms phenomenon – Turbulent pipe flow of dilute polymer solutions*, Mass. Inst. of Tech., (ScD Thesis).

Virk, P. S. (1975). Drag reduction fundamentals. *AICHE Journal, 21*(4), 625–656.

Virk, P. S., Merrill, E. W., Mickley, H. S., Smith, K. A., & Mollo-Christensen, E. L. (1967). The Toms phenomenon: Turbulent pipe flow of dilute polymer solutions. *Journal of Fluid Mechanics, 30*(2), 305–328.

Wells, C. S. (1965). Anomalous turbulent flow of non-newtonian fluids. *AIAA Journal, 3*(10), 1800–1805.

White, A. (1967). *Turbulence and drag reduction with polymer additives.* Research Bulletin No. 4, Hendon College of Technology.

Chapter 3
Velocity Distributions and Boundary-Layer Thicknesses in Turbulent Flows

Curved Tubes

Flow through a curved tube represents a very interesting situation wherein a secondary flow, set up by virtue of the centrifugal force due to curvature, gets superimposed on the axial velocity flow field. The fluid at the center is driven toward the outer wall and then pushed back along the wall toward the inner side, setting up a double motion as shown schematically in Fig. 3.1.

This kind of turbulent flow problem of drag reducing fluids in a curved tube has been studied, and the effect of secondary flow has been determined (Shenoy et al., 1980) for flow through a helical coil.

Coordinate System and Equations of Continuity and Motion

The coordinate system used is shown in Fig. 3.2, where a and $\overline{R}$ are the radii of the helical coil and the curvature, respectively. The flow is assumed to be fully developed and occurring under a constant axial pressure gradient in the direction of increasing Φ.

A drag reducing fluid is considered to be flowing through the coil under turbulent conditions. The fluid is assumed to be dilute enough to maintain its flow behavior index $n = 1$. Its mild elasticity is characterized by the dimensionless Deborah number De. When De $= 0$, the fluid is purely Newtonian, and the results thus obtained can be compared with the work for Newtonian fluids (Ito, 1956, 1959a, 1959b).

A. Shenoy, *Rheology of Drag Reducing Fluids*,
https://doi.org/10.1007/978-3-030-40045-3_3

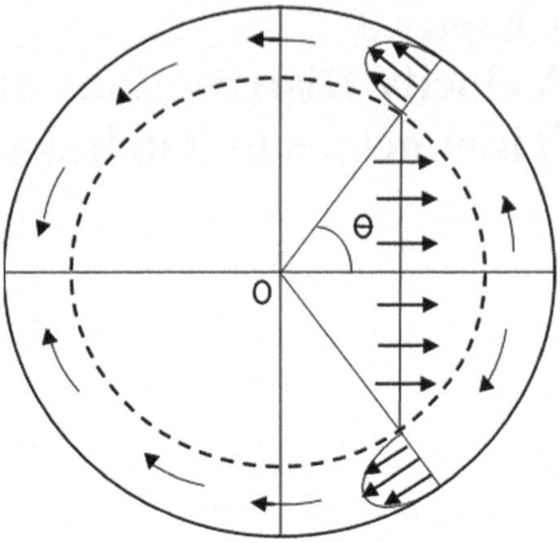

Fig. 3.1 Schematic representation of the flow model for a curved tube showing the boundary layer, inviscid core, and the continuity of secondary flow (Shenoy, Ranade, & Ulbrecht, 1980). (Reprinted with permission from Taylor & Francis and Copyright Clearance Center through RightsLink)

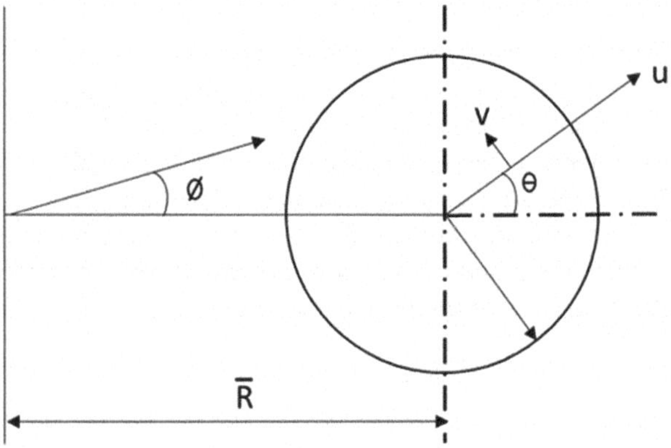

Fig. 3.2 Coordinate system for flow through a helical coil (Shenoy et al., 1980). (Reprinted with permission from Taylor & Francis and Copyright Clearance Center through RightsLink)

For the flow under consideration, the time-averaged equations of continuity and motion are written in their simplified forms as follows:

$$\frac{\partial u}{\partial r}+\frac{u}{r}+\frac{1}{r}\frac{\partial v}{\partial \theta}=0 \tag{3.1}$$

$$\rho\left(u\frac{\partial u}{\partial r}+v\frac{\partial u}{\partial \theta}-\frac{v^2}{r}-\frac{w^2\cos\theta}{\overline{R}}\right)=-\frac{\partial p}{\partial r}+\frac{\partial \tau_{\mathrm{rr}}}{\partial r}+\frac{\tau_{\mathrm{rr}}-\tau_{\theta\theta}}{r}+\frac{1}{r}\frac{\partial \tau_{\mathrm{r}\theta}}{\partial \theta} \tag{3.2}$$

$$\rho\left(u\frac{\partial v}{\partial r}+\frac{v}{r}\frac{\partial v}{\partial \theta}+\frac{uv}{r}+\frac{w^2\sin\theta}{\overline{R}}\right)=-\frac{1}{r}\frac{\partial p}{\partial \theta}+\frac{\partial \tau_{\mathrm{r}\theta}}{\partial r}+\frac{1}{r}\frac{\partial \tau_{\theta\theta}}{\partial \theta}+\frac{2\tau_{\mathrm{r}\theta}}{r} \tag{3.3}$$

$$\rho\left(u\frac{\partial w}{\partial r}+\frac{v}{r}\frac{\partial w}{\partial \theta}\right)=-\frac{1}{\overline{\overline{R}}}\frac{\partial p}{\partial Ø}+\frac{\partial \tau_{\mathrm{r}Ø}}{\partial r}+\frac{1}{r}\frac{\partial \tau_{\theta Ø}}{\partial \theta}+\frac{2\tau_{\mathrm{r}Ø}\sin\theta}{\overline{R}} \tag{3.4}$$

with the constant axial pressure gradient given as

$$-\frac{\partial p}{\partial \varnothing} = \overline{C} \tag{3.5}$$

Flow Model

The overall flow is considered to be representable by a two-region model consisting of a central inviscid core and a thin boundary layer near the wall as shown in Fig. 3.1. The theoretical analysis is performed for the central inviscid core and the outer thin boundary layer separately.

Equations of Motion for the Inviscid Core

In the core region, it is assumed that the viscous stresses and the Reynolds stresses are negligible and the predominant stresses are only those caused by the secondary flow. It is also assumed that the secondary velocity components u and v are small in comparison with w in this region. Under these assumptions, the equations of motion in the inviscid core reduce to

$$-\frac{w^2 \cos\theta}{\overline{R}} = -\frac{1}{\rho}\frac{\partial p}{\partial r} \tag{3.6}$$

$$\frac{w^2 \sin\theta}{\overline{R}} = -\frac{1}{\rho r}\frac{\partial p}{\partial \theta} \tag{3.7}$$

$$u\frac{\partial w}{\partial r} + \frac{v}{r}\frac{\partial w}{\partial \theta} = \frac{\overline{C}}{\rho \overline{R}} \tag{3.8}$$

A stream function ψ is introduced such that

$$u = \frac{1}{r}\frac{\partial \psi}{\partial \theta}; v = -\frac{\partial \psi}{\partial r} \tag{3.9}$$

Knowing that

$$\frac{\partial}{\partial y} = \sin\theta\frac{\partial}{\partial r} + \frac{\cos\theta}{r}\frac{\partial}{\partial \theta} \tag{3.10}$$

p is eliminated from Eqs. (3.6) and (3.7) to give

$$\frac{\partial w}{\partial y} = 0 \tag{3.11}$$

Thus,

$$w = F(x) \tag{3.12}$$

and can be written as

$$w = \overline{A} + \overline{B}\, r \cos\theta \tag{3.13}$$

Using Eqs. (3.8), (3.9), and (3.13), an expression for ψ can be obtained as

$$\psi = -\frac{\overline{C}}{\rho R\overline{B}}\, r \sin\theta \tag{3.14}$$

Thus, from Eq. (3.9) it is seen that

$$u = \frac{\overline{C}}{\rho R\overline{B}} \cos\theta \tag{3.15}$$

$$v = -\frac{\overline{C}}{\rho R\overline{B}} \sin\theta \tag{3.16}$$

Imposing the condition that at $r = 0$, $w = v_{\mathrm{m}}$ (the mean axial velocity), it can be seen that $\overline{A} = v_{\mathrm{m}}$ from Eq. (3.13). The constant $\overline{B}$ is determined using the condition of continuity of secondary flow as follows: The flux from left to right across the curve ACB drawn outside the boundary layer in Fig. 3.1 is used in the determination of $\overline{B}$

$$\int_0^{\delta} v d\overline{\xi} = \frac{\overline{C}}{\rho R\overline{B}} (\overline{a} - \delta) \sin\theta \tag{3.17}$$

where

$$\overline{\xi} = \overline{a} - r \tag{3.18}$$

under the assumption of a thin boundary layer, Eq. (3.17) can be simplified to give

$$\overline{B} = \frac{\overline{C}\overline{a} \sin\theta}{\rho \overline{R} \int_0^{\delta} v d\overline{\xi}} \tag{3.19}$$

Integrating over the entire cross section of the tube gives

$$\overline{B} = \frac{\overline{C}\overline{a}}{\rho \overline{R}} \frac{1}{\pi} \int_0^{\pi} \frac{\sin\theta d\theta}{\int_0^{\delta} v d\overline{\xi}} \tag{3.20}$$

Boundary-Layer Equations

The axial velocity component w falls sharply to zero at the wall and hence within the boundary layer w becomes comparable to v. An order of magnitude analysis of Eqs. (3.1), (3.2), (3.3), and (3.4) is now performed using the following:

$$u \sim \mathrm{O}(\delta);\quad v, w \sim \mathrm{O}(1);\quad \frac{\partial}{\partial r} \sim \mathrm{O}\left(\frac{1}{\delta}\right);\quad \frac{\partial}{\partial \theta}, \frac{\partial}{\partial \varnothing} \sim \mathrm{O}(1) \tag{3.21}$$

where δ is the boundary-layer thickness.

The simplified forms of the boundary-layer equations are then written as follows:

$$\frac{\partial u}{\partial r} + \frac{1}{\overline{a}}\frac{\partial v}{\partial \theta} = 0 \tag{3.22}$$

$$\frac{v^2}{\overline{a}} + \frac{w^2 \cos\theta}{\overline{R}} = 0 \tag{3.23}$$

$$u\frac{\partial v}{\partial r} + \frac{v}{\overline{a}}\frac{\partial v}{\partial \theta} = \frac{w_1^2 - w^2}{\overline{R}}\sin\theta + \frac{1}{\rho}\frac{\partial \tau_{r\theta}}{\partial r} \tag{3.24}$$

$$u\frac{\partial w}{\partial r} + \frac{v}{\overline{a}}\frac{\partial w}{\partial \theta} = \frac{\overline{C}}{\rho \overline{R}} + \frac{1}{\rho}\frac{\partial \tau_{r\varnothing}}{\partial r} \tag{3.25}$$

Note that in Eq. (3.23), $\partial p/\partial r$ is neglected, and in Eq. (3.24), $\partial p/\partial \theta$ is substituted from Eq. (3.7) as

$$\frac{\partial p}{\partial \theta} = -\frac{\rho \overline{a}}{\overline{R}} w_1^2 \sin\theta \tag{3.26}$$

where w_1 is the axial velocity at the edge of the boundary layer.

$\overline{C}$ in Eq. (3.25) is calculated by considering the balance between the axial pressure gradient and the frictional stress at the wall to give

$$\overline{C} = \frac{2\overline{R}}{\pi \overline{a}} \int_0^{\pi} \tau_{r\varnothing}\Big|_{r=\overline{a}} d\theta \tag{3.27}$$

From the explicit expression for the wall shear stress τ_w to be obtained later in Eq. (3.45), it will be obvious that, for the dilute drag reducing fluids ($n = 1$),

$$\tau_w \sim \rho v_m^2 (\mathrm{Re})^{-\beta} \left(\frac{\delta}{\overline{a}}\right)^{-\beta} \tag{3.28}$$

where Re is the Reynolds number and β is a function of the Deborah number De derived from the Blasius-type friction factor equation.

A further order of magnitude analysis of Eqs. (3.22), (3.23), (3.24), and (3.25) is performed using $v \sim \epsilon v_{\mathrm{m}}$ and $w \sim v_{\mathrm{m}}$ where ϵ is a small quantity, to obtain

$$\frac{\delta}{\overline{a}} \sim \left[\mathrm{Re} \left(\frac{\overline{a}}{\overline{R}} \right)^{\frac{1}{2\beta}} \right]^{-\frac{\beta}{\beta+1}} \tag{3.29}$$

where

$$\mathrm{Re} = \frac{(2\overline{a}) v_{\mathrm{m}} \rho}{\mu} \tag{3.30}$$

$$\overline{C} \sim \rho v_{\mathrm{m}}^2 (\mathrm{Re})^{-\frac{\beta}{\beta+1}} \left(\frac{\overline{a}}{\overline{R}} \right)^{\frac{\beta}{2(\beta+1)}} \tag{3.31}$$

and consequently

$$\frac{\overline{C}}{\rho \overline{R}} \sim \mathrm{O}\left(\frac{\delta}{\overline{a}} \right) \tag{3.32}$$

For $\delta/\overline{a} \ll 1$, at $r = \overline{a} - \delta$

$$v \cong 0; \quad \frac{\partial v}{\partial r} \cong 0 \tag{3.33}$$

$$w = w_1; \quad \frac{\partial w}{\partial r} \cong 0 \tag{3.34}$$

A further boundary condition is the no-slip at the tube wall which gives

$$v = w = 0 \tag{3.35}$$

Momentum Integral Equations of the Boundary Layer

The integration of the equations of motion (3.14) and (3.15) within the boundary layer, and their simplifications using the conditions developed in the foregoing, yield

$$\tau_{\mathrm{r}\theta} \frac{\overline{a}}{\rho} \bigg|_{\overline{\xi}=0} = -\frac{d}{d\theta} \int_0^{\delta} v^2 d\overline{\xi} + \frac{\overline{a}}{\overline{R}} \sin\theta \int_0^{\delta} \left(w_1^2 - w^2 \right) d\overline{\xi} \tag{3.36}$$

$$\tau_{r\emptyset}\frac{\overline{a}}{\rho}\bigg|_{\overline{\xi}=0} = w_1\frac{d}{d\theta}\int_0^\delta v d\overline{\xi} - \frac{d}{d\theta}\int_0^\delta vw d\overline{\xi} \tag{3.37}$$

where $\overline{\xi} = \overline{a} - r$ and, consequently, the conditions $\overline{\xi} = 0$ and $\overline{\xi} = \delta$ correspond to the tube wall and the edge of the boundary layer, respectively.

Solution of the Momentum Integral Equations

In keeping with the concept of the integral technique, a suitable form for the axial and angular velocity distributions in the boundary layer needs to be assumed to obtain the solutions of the momentum integral equations. The forms for the velocity distributions are sought by first obtaining expressions for the local shear stress and the mean axial velocity distribution in the turbulent boundary-layer flow over a flat plate of zero angle of incidence and then extending it for the turbulent boundary-layer flow at hand.

Using the fact that in an approximation, Eq. (2.27) given in Chap. 2 is valid at any distance y rather than at the wall specifically, the turbulent velocity profile in the tube flow can be expressed in the following form:

$$\frac{\overline{u}}{\overline{u}_m} = \left(\frac{y}{\overline{a}}\right)^{\frac{\beta}{(2-\beta)}} \tag{3.38}$$

This can be further modified for the turbulent boundary layer over a flat plate with zero angle of incidence as

$$\frac{u_1}{u_e} = \left(\frac{y}{\delta}\right)^{\frac{\beta}{(2-\beta)}} \tag{3.39}$$

where δ is the turbulent boundary-layer thickness and u_e is the velocity at the edge of the boundary layer. The expression for the mean wall shear stress for dilute drag reducing fluids as given by Eq. (2.33) in Chap. 2 is written as follows:

$$\tau_w = \frac{\alpha}{2^{1+\beta}}\left[\frac{(2-\beta)^2}{(4-\beta)}\right]^{2-\beta} \rho u_e^{2-\beta}\left[\frac{\mu}{\rho\delta}\right]^\beta \tag{3.40}$$

Thus, Eqs. (3.39) and (3.40) represent the local velocity distribution and the local shear stress in the turbulent boundary-layer flow over a flat plate at zero angle of incidence. It now remains to make use of these distributions to develop explicit expressions for the turbulent boundary-layer flow at hand.

The angular and axial velocity distributions in the boundary layer are assumed to have the following form based on boundary conditions and continuity of flow:

$$v = D_1 \left(\frac{\xi}{\delta}\right)^{\frac{\beta}{(2-\beta)}} \left(1 - \frac{\xi}{\delta}\right) \tag{3.41}$$

$$w = w_1 \left(\frac{\xi}{\delta}\right)^{\frac{\beta}{(2-\beta)}} \tag{3.42}$$

The resulting velocity in the neighborhood of the wall surface ($\xi \to 0$) can be found as

$$\lim_{\xi \to 0} \sqrt{v^2 + w^2} = U_1 \left(\frac{\xi}{\delta}\right)^{\frac{\beta}{(2-\beta)}} \tag{3.43}$$

where

$$U_1 = \sqrt{D_1^2 + w_1^2} \tag{3.44}$$

Assuming that Eq. (3.40) will be valid even if u_e is replaced by U_1 yields

$$\tau_w = \frac{\alpha}{2^{1+\beta}} \left[\frac{(2-\beta)^2}{(4-\beta)}\right]^{2-\beta} \rho U_1^{\,2-\beta} \left[\frac{\mu}{\rho\delta}\right]^{\beta} \tag{3.45}$$

Noting that

$$\frac{(\tau_{r\theta})_{\xi \to 0}}{(\tau_{r\emptyset})_{\xi \to 0}} = \lim_{\xi \to 0} \frac{v}{w} = \frac{D_1}{w_1} \tag{3.46}$$

and

$$\tau_w = \sqrt{(\tau_{r\theta})^2_{\xi \to 0} + (\tau_{r\emptyset})^2_{\xi \to 0}} \tag{3.47}$$

gives

$$\tau_{r\theta} = \frac{D_1}{\sqrt{D_1^2 + w_1^2}} \tau_w \tag{3.48}$$

$$\tau_{r\emptyset} = \frac{w_1}{\sqrt{D_1^2 + w_1^2}} \tau_w \tag{3.49}$$

An expression for w_1 can be obtained from Eq. (3.13) using the assumption of a thin boundary layer as

$$w_1 = v_m + \overline{B}\,\overline{a}\,\cos\theta \tag{3.50}$$

The following expressions are used for nondimensionalizing:

$$\delta = \delta_c \overline{a}\left\{\left(\mathrm{Re}\,\frac{\overline{a}}{\overline{R}}\right)^{\frac{1}{2\beta}}\right\}^{\frac{-\beta}{(\beta+1)}} \tag{3.51}$$

$$D_1 = D_c\left\{v_m\left(\frac{\overline{a}}{\overline{R}}\right)^{\frac{1}{2}}\right\} \tag{3.52}$$

$$w_1 = w_c(v_m) \tag{3.53}$$

Equations (3.24) and (3.25) then reduce to

$$f_1\frac{\partial}{\partial\theta}\left[D_c^2\delta_c\right] = \left[1 - f_2\right]\delta_c w_c^2 \sin\theta - f_3 2^{\beta}\delta_c{}^{-\beta} D_c w_c^{1-\beta} \tag{3.54}$$

$$f_4\frac{\partial}{\partial\theta}\left[D_c w_c \delta_c\right] - f_5 w_c \frac{\partial}{\partial\theta}\left[D_c\delta_c\right] = -f_3 2^{\beta}\delta_c{}^{-\beta} w_c^{1-\beta} \tag{3.55}$$

The nondimensional form of Eq. (3.50) then becomes

$$w_c = 1 + \frac{\overline{B}\,\overline{a}}{v_m}\cos\theta \tag{3.56}$$

where from Eq. (3.20)

$$\frac{B_o\,\overline{a}}{v_m} = \frac{\alpha^*\pi^{-1}}{f_4}\int_0^{\pi}\frac{\sin\theta}{D_c\delta_c}d\theta \tag{3.57}$$

and

$$\alpha^* = f_5 2^{\beta+1}\pi^{-1}\int_0^{\pi} w_c^{2-\beta}\delta_c^{-\beta}d\theta \tag{3.58}$$

The functions f_1 to f_5 are given as

$$f_1 = \frac{(2-\beta)^3}{2(2+\beta)(6-\beta)} \tag{3.59}$$

$$f_2 = \frac{(2-\beta)}{(2+\beta)} \tag{3.60}$$

$$f_3 = \frac{\alpha\left[\frac{(2-\beta)^2}{(4-\beta)}\right]^{2-\beta}}{2^{\beta+1}} \tag{3.61}$$

$$f_4 = \frac{(2-\beta)^2}{2(4-\beta)} \tag{3.62}$$

$$f_5 = \frac{(2-\beta)^2}{4(2+\beta)} \tag{3.63}$$

Since α and β are functions of Deborah number, f_1 to f_5 are all functions of De alone. The values of f_1 to f_5 for varying Deborah numbers are reported in Table 3.1.

The relationship for friction factor can be obtained by using Eq. (3.27) along with the definition of the average friction factor:

$$f_c = \frac{4\left(-\frac{\partial p}{\overline{\overline{R}}\partial \varnothing}\right)\overline{a}}{\left[\mathrm{Re}\left(\frac{\overline{a}}{\overline{\overline{R}}}\right)^{\frac{1}{2\beta}}\right]^{\frac{\beta}{\beta+1}}} \tag{3.64}$$

which on proper substitution gives

$$f_c = \frac{4\alpha^*\left(\frac{\overline{a}}{\overline{\overline{R}}}\right)^{\frac{1}{2}}}{\left[\mathrm{Re}\left(\frac{\overline{a}}{\overline{\overline{R}}}\right)^{\frac{1}{2\beta}}\right]^{\frac{\beta}{\beta+1}}} \tag{3.65}$$

Thus, the appropriate similarity variable in the case of turbulent flow of dilute drag reducing fluids turns out to be $\mathrm{Re}\left(\frac{\overline{a}}{\overline{\overline{R}}}\right)^{\frac{1}{2\beta}}$ and can be referred to as the modified Dean number Dn.

Method of Solution

Equations (3.54), (3.55), (3.56), (3.57), (3.58), (3.59), (3.60), (3.61), (3.62), and (3.63) form a set of complex equations which cannot be solved analytically, and numerical techniques must be resorted to. The initial values of δ_c, D_c, and w_c needed to start the integration are obtained by expanding in the neighborhood of $\theta = 0$ as follows:

Table 3.1 Values of various parameters for flow of dilute drag reducing fluids through curved tubes (Shenoy et al., 1980)

De	α	β	f_1	f_2	f_3	f_4	f_5	δ_c	$\alpha*$	$\frac{\overline{B}\,\overline{a}}{v_m}$
0	0.0790	0.25	0.2071	0.875	0.02332	0.4083	0.3403	0.6057	0.0719	0.3793
Ito								0.6060	0.0720	0.3800
1	0.06671	0.248	0.2079	0.7793	0.01971	0.4090	0.3413	0.5486	0.0644	0.3776
2	0.05784	0.2456	0.2089	0.7813	0.01711	0.4099	0.3427	0.4939	0.0572	0.3755
3	0.05143	0.2448	0.2093	0.7819	0.01522	0.4102	0.3431	0.4510	0.0519	0.3747
4	0.04677	0.2453	0.2091	0.7815	0.01384	0.4100	0.3428	0.4170	0.0482	0.3752
5	0.04377	0.248	0.2080	0.7794	0.01295	0.4091	0.3414	0.3914	0.0459	0.3776
6	0.04119	0.2507	0.2069	0.7773	0.01215	0.4081	0.3399	0.3693	0.0440	0.3780
7	0.03967	0.255	0.2051	0.7739	0.01168	0.4066	0.3376	0.3531	0.0431	0.3836
8	0.03883	0.2603	0.2029	0.7696	0.01141	0.4046	0.3347	0.3412	0.0429	0.3882
9	0.03855	0.2666	0.2004	0.7647	0.01129	0.4024	0.3314	0.3326	0.0433	0.3935
10	0.03863	0.2734	0.1977	0.7594	0.01126	0.3999	0.3278	0.3260	0.0439	0.3990

$$\delta_c = \delta_1\left(1 + \delta_2\theta^2 + \cdots\right) \tag{3.66}$$

$$D_c = D_2\theta\left(1 + D_3\theta^3 + \cdots\right) \tag{3.67}$$

$$w_c = 1 + \frac{\overline{B}\,\overline{a}}{v_m}\left(1 - \frac{\theta^2}{2!} + \frac{\theta^4}{4!} - \cdots\right) \tag{3.68}$$

Substituting these into Eqs. (3.41) and (3.42) and rearranging and equating equal powers of δ_c yield

$$\delta_1 = [f_5]^{\frac{1}{(1+\beta)}} 2^{\frac{\beta}{(1+\beta)}} \left[\frac{2f_1}{(f_3 - f_4)^2} - \frac{1}{f_3 - f_4}\right]^{\frac{1}{2(1+\beta)}} [1 - f_2]^{-\frac{1}{2(1+\beta)}} \left[1 + \frac{\overline{B}\overline{a}}{v_m}\right]^{-\frac{\beta}{(1+\beta)}} \tag{3.69}$$

$$D_1 = \left[-\frac{f_5}{f_3 - f_4}\right] 2^{\beta} \left[1 + \frac{\overline{B}\overline{a}}{v_m}\right]^{1-\beta} \delta_c^{-(1-\beta)} \tag{3.70}$$

To start the integration procedure, the value of $\frac{\overline{B}\overline{a}}{v_m}$ was first assumed so that δ_1 is known. The numerical integration is then started for obtaining $\delta_c(\theta)$ and $D_c(\theta)$ from Eqs. (3.54) and (3.55) using a Runge-Kutta-Merson technique. Substitution of these $\delta_c(\theta)$ and $D_c(\theta)$ into Eq. (3.58) gives a value of $\alpha*$ which is in turn used to calculate $\frac{\overline{B}\overline{a}}{v_m}$ from Eq. (3.57). The obtained value is compared with the initial assumed value, and the computations are continued until the difference between the two values is less than 10^{-4}. The calculations are performed for $0 \leq \mathrm{De} \leq 10$.

Discussion of the Numerical Results

Table 3.1 gives the values of the functions α, β, and f_1 to f_5 and the computed values of δ_1, $\alpha*$, and $\frac{\overline{B}\overline{a}}{v_m}$ for dilute drag reducing fluids with $0 \leq \mathrm{De} \leq 10$. The excellent agreement between the computed values of δ_1, $\alpha*$, and $\frac{\overline{B}\overline{a}}{v_m}$ for $\mathrm{De} = 0$ and those obtained for a Newtonian fluid (Ito, 1959a, 1959b) gives evidence of the propriety of the numerical scheme.

The numerical results of δ_c, w_c, and D_c are plotted as functions of θ in Figs. 3.3, 3.4, and 3.5, with the varying parameter as De.

It is evident from Figs. 3.4 and 3.5 that the dimensionless axial velocity component w_c and the dimensionless characteristic angular velocity component D_c do not change significantly with increasing De. However, Fig. 3.3 shows that there is a marked decreasing trend in the dimensionless boundary-layer thickness δ_c with increasing De. This is clearly in contradiction with one of the models which aspires to explain the mechanism of drag reduction, namely, the thickening of the boundary layer. This discrepancy does, in turn, lend more support to alternative mechanisms of drag reduction, such as the reduced rate of generation of turbulent eddies or the

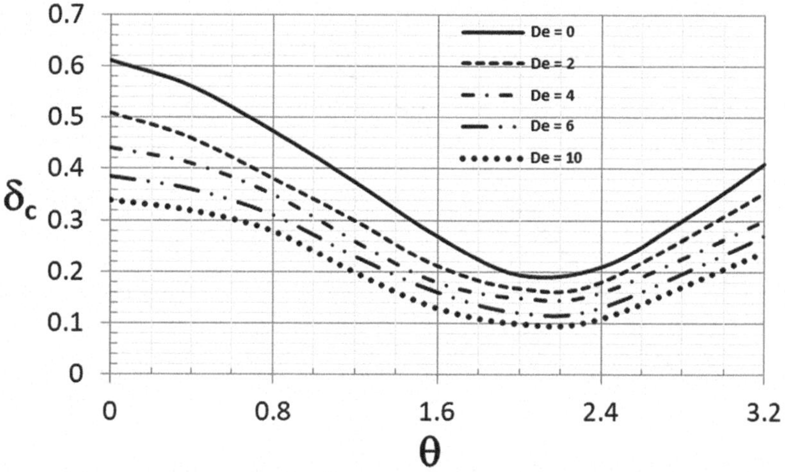

Fig. 3.3 Variation of boundary-layer thickness δ_c with θ and De (Shenoy et al., 1980). (Reprinted with permission from Taylor & Francis and Copyright Clearance Center through RightsLink)

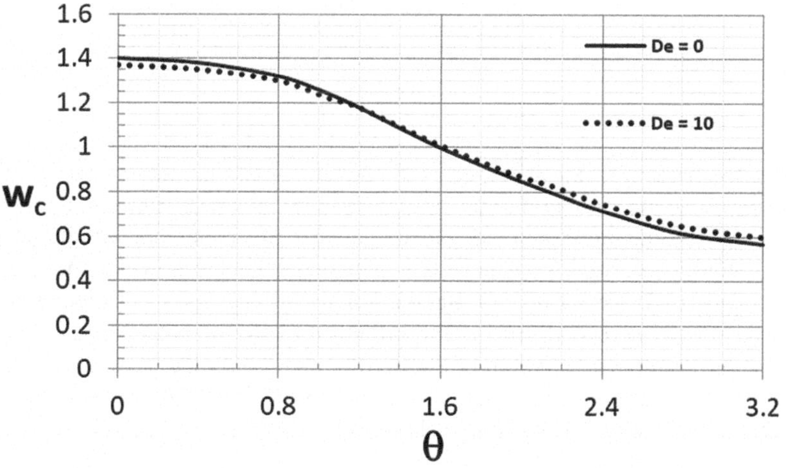

Fig. 3.4 Variation of axial velocity w_c with θ and De (Shenoy et al., 1980). (Reprinted with permission from Taylor & Francis and Copyright Clearance Center through RightsLink)

reduced intensity of energy dissipation in the bulk flow. In any case, Figs. 3.3, 3.4, and 3.5, in conjunction with Eqs. (3.51), (3.52), and (3.53) and Eq. (3.13), provide a complete velocity distribution for the turbulent flow of a mildly viscoelastic (drag reducing) liquid through a curved tube.

Finally, the dependence of the friction factor on both the process and the material parameters needs to be discussed. It follows from Eqs. (3.58) and (3.65) that

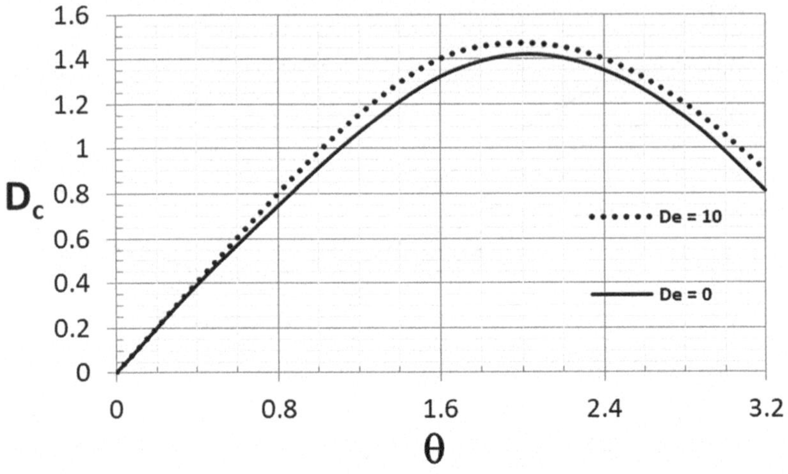

Fig. 3.5 Variation of angular velocity D_c with θ and De (Shenoy et al., 1980). (Reprinted with permission from Taylor & Francis and Copyright Clearance Center through RightsLink)

$$f_c = \frac{4\left(\frac{\overline{a}}{R}\right)^{\frac{1}{2}} f_5 2^{\beta+1} \pi^{-1} \int_0^{\pi} w_c^{2-\beta} \delta_c^{-\beta} d\theta}{\left[\mathrm{Re}\left(\frac{\overline{a}}{R}\right)^{\frac{1}{2\beta}}\right]^{\frac{\beta}{\beta+1}}} \tag{3.71}$$

Thus, the friction factor increases as the dimensionless boundary-layer thickness decreases. Since the reduction of the boundary-layer thickness runs parallel with the increase of Deborah number, it may be concluded erroneously that with the increase of Deborah number, friction factor will also increase. This is, however, not the case. An inspection of Table 3.1 reveals that the function f_5 decreases as Deborah number increases, which more than offsets the influence of the boundary-layer thickness. Thus, the net result is a reduction of the friction factor with the increase of De. This reduction can be now easily predicted using Eq. (3.65) and the corresponding values of α (De), β (De), and $a*$(De) from Table 3.1.

Comparison of Theoretical Results with Experimental Data

Comparison of the results of the theoretical analysis with experimental data is made possible by reported data (Devarajan, 1976). Pressure drop was measured for several dilute drag reducing solutions flowing through four helical coils of dimensionless radii of curvature $\frac{\overline{a}}{R}$ ranging from 0.0097 to 0.135. The obvious difficulty in estimating the Deborah numbers for the drag reducing fluids limits the comparison between theory and experiments to 100 ppm polyacrylamide (PAA) solution alone.

For 100 ppm PAA solution, the Deborah number is estimated (Seyer & Metzner, 1969) to be 5.8 for a wall shear stress of 4.788×10^2 dynes/cm^2. Using experimental data (Oliver, 1966), it is evident that for the 100 ppm PAA solutions considered herein

$$\mathrm{De} \propto \sqrt{\tau_{\mathrm{w}}} \tag{3.72}$$

An estimate of the shear stress for turbulent flow of 100 ppm PAA solution through a helical coil can be made as follows:

$$\tau_{\mathrm{w}} \cong \frac{\alpha(0.817)^{2-\beta}}{2} \rho v_{\mathrm{m}}^2 (\,Re\,)^{\frac{-\beta}{\beta+1}} \left(\frac{\overline{a}}{\overline{R}}\right)^{\frac{-\beta}{2(\beta+1)}} \tag{3.73}$$

Using values of α and β as that for water and Re $= 21000$, $v_{\mathrm{m}} = 111$ cm/s, and $\frac{\overline{a}}{\overline{R}} = 0.135$, the shear stress $\tau_{\mathrm{w}} \cong 57$ dynes/cm^2. It is seen that this estimated value of τ_{w} is quite insensitive to the value of $\frac{\overline{a}}{\overline{R}}$. The changes in v_{m} also do not seem to affect it drastically as the term $(\,\mathrm{Re}\,)^{\frac{-\beta}{\beta+1}}$ more or less compensates the variation.

Therefore, it would be reasonable to assume that the estimated value of τ_{w} is valid for the 100 ppm PAA solution over the $\frac{\overline{a}}{\overline{R}}$ values and the range of Reynolds numbers in the experimental data. Using Eq. (3.73) and the proposed value of De $= 5.8$ at $\tau_{\mathrm{w}} = 4.788 \times 10^2$dynes/cm^2 (Seyer & Metzner, 1969), estimated value of De for turbulent flow of 100 ppm PAA solution through helical coils approximately equals 2.

For De $= 2$ the corresponding value of $\alpha*$ is taken from Table 3.1, and the theoretical friction factor values are obtained using Eq. (3.65). For each coil with fixed $\frac{\overline{a}}{\overline{R}}$, the theoretical friction factors are plotted versus the experimental friction factors (Devarajan, 1976) in the fully turbulent region. Table 3.2 shows the comparison, and the good agreement between the theory and experiments, at least in the range of verification, merits the recommendation of the use of Eq. (3.65) in design practice.

At a given flow rate, the turbulent flow of mildly viscoelastic liquids through curved tubes shows a lower pressure drop than one would expect for an inelastic shear-thinning

Table 3.2 Predicted friction factor (f_{p}) with experimental values (f_{e}) for 100 ppm PAA solution with estimated De $= 2$; $\beta = 0.2456$; $\alpha* = 0.05715$ (Shenoy et al., 1980)

$\frac{\overline{a}}{\overline{R}} \rightarrow$	0.0097		0.0188		0.0463		0.135	
↓Re	f_{p}	f_{e}	f_{p}	f_{e}	f_{p}	f_{e}	f_{p}	f_{e}
10180	0.0235	0.0232	0.0250	0.0278	0.0274	0.0292	0.0304	0.0352
11230	0.0230	0.0200	0.0246	0.0274	0.0268	0.0296	0.0298	0.0344
12280	0.0226	0.0184	0.0241	0.0266	0.0264	0.0280	0.0293	0.0328
21000	–	–	0.0217	0.0241	0.0237	0.0264	0.0264	0.0304
34450	–	–	–	–	0.0215	0.0216	0.0239	0.0264
39300	–	–	0.0209	0.0210	0.0209	0.0214	0.0233	0.0256

liquid having the same flow curve. The above analysis leads to an equation for the calculation of the friction factor as a function of Deborah number, the dimensionless measure of the liquid's viscoelasticity. The predictive value of Eq. (3.65) when tested against some published experimental data is found to be good. As an offshoot of this analysis, it is concluded that the increase of Deborah number, although leading to a reduction of the friction factor, does not lead to a thickening of the boundary layer.

Rotating Straight Circular Tubes

When a fluid flows through a straight tube rotating at a constant angular velocity about an axis perpendicular to its own, a secondary flow is set up by virtue of the Coriolis force arising from the rotation, and this superimposition of the secondary flow field upon the axial velocity field has interesting implications. For example, an important technical application of such flows is in a rotating power transmission shaft that is longitudinally bored and through which a fluid is pumped for cooling.

The flow of Newtonian fluid through a straight tube rotating at a constant angular velocity about an axis perpendicular to its own has been extensively studied as can be seen in Table 3.3. In contrast, however, there exists only one study of the turbulent flow of viscoelastic fluids in such flow situations (Shenoy, 1986). A theoretical study of the turbulent flow of mildly elastic drag reducing fluids in a tube rotating around an axis perpendicular to its own at high Reynolds numbers and in the range of high rotational parameters has been done. This range permits the assumption of a frictionless central core surrounded by a thin boundary layer so that an approximate boundary-layer analysis can be used.

Table 3.3 Theoretical studies on the flow of Newtonian fluids through a tube rotating about an axis perpendicular to its own (Shenoy, 1986)

No.	Authors	Remarks
1	Barua (1963)	Series solution, first approximation
2	Benton (1956)	Secondary flow analysis
3	Benton and Boyer (1966)	Perturbation technique solution, for small value of angular rotation
4	Gunn, Mena, and Walters (1974)	Perturbation technique solution, second approximation
5	Ito and Nanbu (1971)	Laminar boundary-layer analysis; Polhausen's method
6	Mori and Nakayama (1968)	Fully developed flow field and temperature field; boundary-layer approximation
7	Nanbu (1970)	Turbulent boundary-layer analysis
8	Jones, Trevors, and Walters (1967)	Series solution: velocity distribution and Stream functions

Analysis Approach

It is assumed that the flow is fully developed and taking place under a constant axial pressure gradient in the *z*-direction. The analysis is performed along the lines for Newtonian fluid (Nanbu, 1970) and the results checked for the propriety of the analysis.

Flow Model

The overall flow is divided into two regions: a central inviscid core and a thin boundary layer adjacent to the wall as shown by the schematic representation in Fig. 3.6.

The fluid at the center is driven toward the tube wall by the Coriolis force caused due to the rotation of the tube and returns to the center following the tube wall, thus setting up a double vertical motion in the cross section of the pipe. In the thin boundary layer, the velocity is affected by the viscosity, and hence boundary-layer approximations can be applied in the analysis. On the other hand, in the core region, the velocity field is affected mainly by the secondary flow, and the effect of viscosity can be disregarded. Such a flow model is justified because the effect of Coriolis forces is like the effect of centrifugal forces on flow in a curved pipe, and the resulting similarity suggests that abovementioned flow model which has been used with success for analysis in a curved pipe described in the previous section can likewise be adapted to the rotational straight circular pipe situation.

As the envisaged model consists of two regions, different sets of coordinate systems are used to ease the analysis. The Cartesian coordinates are convenient for the core region analysis, while cylindrical coordinates are appropriate for the boundary region (see Fig. 3.7).

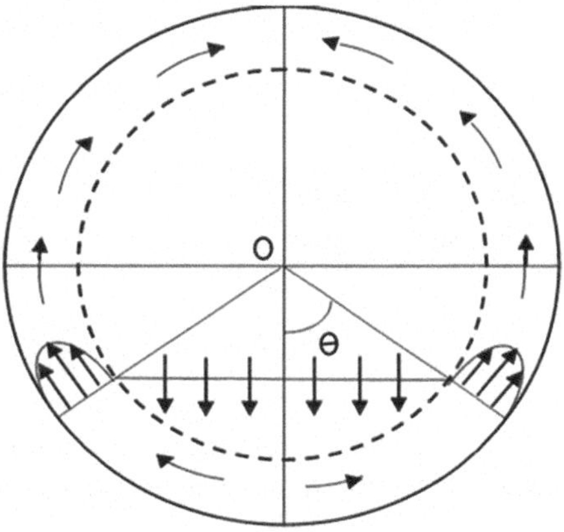

Fig. 3.6 Schematic representation of the flow model for a rotating straight circular tube showing the boundary layer, inviscid core, and the continuity of secondary flow (Shenoy, 1986). (Reprinted with permission from Springer and Copyright Clearance Center through RightsLink)

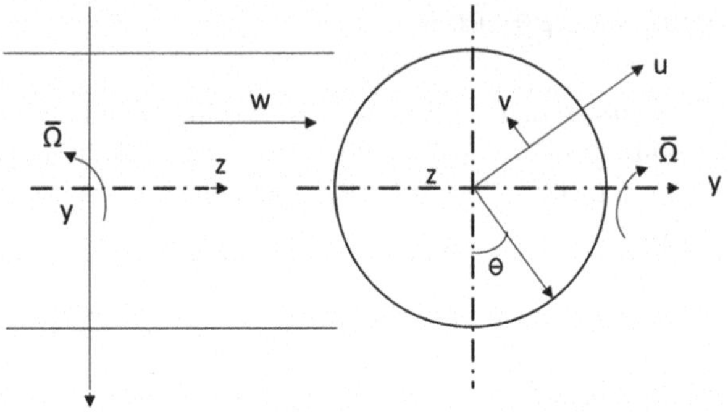

Fig. 3.7 Coordinate system for a rotating straight circular tube (Shenoy, 1986). (Reprinted with permission from Springer and Copyright Clearance Center through RightsLink)

Motion in the Inviscid Core

For the flow in the inviscid core, the simplified forms of the time-averaged equations of motion with velocity components u, v, and w in the x-, y-, and z-directions, respectively, can be written as

$$u\frac{\partial u}{\partial x}+v\frac{\partial u}{\partial y}=-\frac{1}{\rho}\frac{\partial p^*}{\partial x}+2\overline{\Omega}w \tag{3.74}$$

$$u\frac{\partial v}{\partial x}+v\frac{\partial v}{\partial y}=-\frac{1}{\rho}\frac{\partial p^*}{\partial y} \tag{3.75}$$

$$u\frac{\partial w}{\partial x}+v\frac{\partial w}{\partial y}=-\frac{1}{\rho}\frac{\partial p^*}{\partial z}-2\overline{\Omega}w \tag{3.76}$$

where

$$p^*=p-\frac{1}{2}\rho\overline{\Omega}^2\left(x^2+z^2\right) \tag{3.77}$$

From the above equations, it is clear that for fully developed flow

$$-\frac{\partial p}{\partial z}=\overline{C} \tag{3.78}$$

where $\overline{C}$ is a constant.

Further the equation of continuity is given by

$$\frac{\partial u}{\partial x}+\frac{\partial v}{\partial y}=0 \tag{3.79}$$

Since the axial velocity distribution is much more uniform in the core, it is assumed that the transverse velocity components u and v are small in comparison with the axial velocity component w. Hence the Eqs. (3.74), (3.75), and (3.76) reduce to

$$0 = -\frac{1}{\rho}\frac{\partial p^*}{\partial x} + 2\overline{\Omega}w \tag{3.80}$$

$$0 = -\frac{1}{\rho}\frac{\partial p^*}{\partial y} \tag{3.81}$$

$$u\frac{\partial w}{\partial x} + v\frac{\partial w}{\partial y} = -\frac{\overline{C}}{\rho} - 2\overline{\Omega}w \tag{3.82}$$

Introducing a stream function ψ, the equation of continuity can be satisfied by

$$u = \frac{\partial \psi}{\partial y}; \quad v = -\frac{\partial \psi}{\partial x} \tag{3.83}$$

Eliminating $p*$ from Eqs. (3.80) and (3.81) and solving for w give

$$w = F(x) \tag{3.84}$$

where F is an arbitrary function of x.

It is assumed that $F(x)$ has the following form so that

$$w = F(x) = \overline{A} + \overline{B}x \tag{3.85}$$

where $\overline{A}$ and $\overline{B}$ are constants to be determined from the equations of continuity of secondary flow.

Substituting Eqs. (3.83) and (3.85) into Eq. (3.82) and integrating with the fact that the flow is symmetrical about the plane $y = 0$:

$$\psi = \frac{\overline{C}y}{\rho(\overline{B} + 2\overline{\Omega})} + \text{constant} \tag{3.86}$$

It is now assumed that the angular velocity $\overline{\Omega}$ is small compared to $\overline{B}$. This would be reasonable because the axial velocity must be sufficiently greater than the angular velocity for boundary-layer approximations to hold. Thus, the simplified form of Eq. (3.86) can be written as

$$\psi = \frac{\overline{C}y}{\rho\overline{B}} + \text{constant} \tag{3.87}$$

Thus, using Eqs. (3.83) and (3.87), it is evident that

$$u = \frac{\overline{C}}{\rho \overline{B}}; \quad v = 0 \tag{3.88}$$

Considering Fig. 3.6 the flux from left to right across the curve *ACB* drawn outside the boundary layer is equal to

$$\int_A^B d\psi = [\psi]_B - [\psi]_A = \frac{\overline{C}}{\rho \overline{B}}(a - \delta) \sin \theta \tag{3.89}$$

Since $v = 0$ at $\theta = 0$ for reasons of symmetry, the above flux is also equal to the flux through the boundary layer at θ, i.e., $\int_0^\delta v d\overline{\xi}$ where $\overline{\xi} = a - r$. Thus,

$$\int_0^\delta v d\overline{\xi} = \frac{\overline{C}a}{\rho \overline{B}} \sin \theta \tag{3.90}$$

assuming a thin boundary layer ($\delta \ll a$).

Integrating over the entire cross section and simplifying give the expression for $\overline{B}$ as

$$\overline{B} = \frac{\overline{C}a}{\rho} \frac{1}{\pi} \int_0^\pi \frac{\sin \theta d\theta}{\int_0^\delta v d\overline{\xi}} \tag{3.91}$$

Further at $r = 0$, $v = w_{\mathrm{m}}$, and hence from Eq. (3.85), an expression for $\overline{A}$ can be as follows:

$$\overline{A} = w_{\mathrm{m}} \tag{3.92}$$

The constants $\overline{A}$ and $\overline{B}$ are thus determined.

Expression for the Axial Pressure Gradient

Considering the force balance, axial pressure gradient, and the frictional stress at the wall gives

$$2\int_0^\pi \int_0^a \left[p^* - \left(p^* + \frac{\partial p^*}{\partial z} \right) \right] r dr d\theta = 2\int_0^\pi [\zeta_{\mathrm{rz}}]_{r=a} a d\theta \tag{3.93}$$

Thus,

$$C = -\frac{\partial p^*}{\partial z} = \frac{2}{\pi a} \int_0^\pi [\tau_{\mathrm{rz}}]_{r=a} d\theta \tag{3.94}$$

Motion in the Boundary Layer

Let u, v, and w be the velocity components in the r-, θ-, and z-directions, respectively, within the boundary layer. The axial velocity component w falls to zero at the wall in the boundary layer, and hence the angular velocity component v becomes comparable to w. Using the following order of magnitude analysis $u \sim O(\delta)$, $v \sim O(1)$, $w \sim O(1)$, $\frac{\partial}{\partial r} = O(\delta^{-1})$, $\frac{\partial}{\partial \theta} = O(1)$, and $\frac{\partial}{\partial z} = O(1)$, the boundary-layer equations of motion and continuity can be written as

$$-\frac{v^2}{a} = -\frac{1}{\rho}\frac{\partial p^*}{\partial r} + 2\overline{\Omega} w \cos\theta \tag{3.95}$$

$$u\frac{\partial v}{\partial r} + \frac{v}{a}\frac{\partial v}{\partial \theta} = -\frac{1}{\rho a}\frac{\partial p^*}{\partial \theta} - 2\overline{\Omega} w \sin\theta + \frac{1}{\rho}\frac{\partial \tau_{r\theta}}{\partial r} \tag{3.96}$$

$$u\frac{\partial w}{\partial r} + \frac{v}{a}\frac{\partial w}{\partial \theta} = -\frac{1}{\rho}\frac{\partial p^*}{\partial z} - 2\overline{\Omega} v \sin\theta + \frac{1}{\rho}\frac{\partial \tau_{rz}}{\partial r} \tag{3.97}$$

$$\frac{\partial u}{\partial r} + \frac{1}{\rho}\frac{\partial v}{\partial \theta} = 0 \tag{3.98}$$

From Eq. (3.95) it is clear that the variation of $p*$ across the boundary layer is of the order of δ and can be neglected.

Using Eqs. (3.80) and (3.81), the pressure gradient in the θ-direction just outside the boundary layer is given by

$$\frac{\partial p^*}{\partial \theta} = -2\rho a\overline{\Omega} w_1 \sin\theta \tag{3.99}$$

where the subscript 1 denotes the conditions at the edge of the boundary layer.

From the turbulent boundary analysis (Shenoy et al., 1980), it can be shown that the wall shear stress τ_w would vary as follows:

$$\tau_w \sim \rho w_m^2 (\,\mathrm{Re}\,)^{-\beta}\left(\frac{\delta}{a}\right)^{-\beta} \tag{3.100}$$

where for dilute drag reducing fluids β is a function of Deborah number $\mathrm{De} = (\theta_{fl} u*^2)/\nu$ derived from the Blasius-type friction factor (Shenoy & Mashelkar, 1983) developed for drag reducing fluids. The Reynolds number is defined as follows:

$$\mathrm{Re} = \frac{2 a w_m \rho}{\mu} \tag{3.101}$$

A further order of magnitude analysis of Eqs. (3.95), (3.96), (3.97), and (3.98) is performed using $v \sim \epsilon w_m$ and $w \sim w_m$ where ϵ is a small quantity, thereby giving

$$\frac{\delta}{a} \sim \left[\mathrm{Re} \left(\frac{2a\Omega}{w_{\mathrm{m}}} \right)^{\frac{1}{2\beta}} \right]^{\frac{-\beta}{\beta+1}} \tag{3.102}$$

and

$$C \sim \rho w_{\mathrm{m}}^{2} (\,\mathrm{Re}\,)^{\frac{-\beta}{(\beta+1)}} \left(\frac{2a\Omega}{w_{\mathrm{m}}} \right)^{\frac{\beta}{2(\beta+1)}} \tag{3.103}$$

Consequently

$$C \sim \mathrm{O}\left(\frac{\delta}{a}\right) \tag{3.104}$$

Following the assumptions that $\delta \ll a$, the following would hold approximately at the edge of the boundary layer, i.e., at $r = a - \delta$.

$$v = 0, \quad \frac{\partial v}{\partial r} = 0 \tag{3.105}$$

$$w = w_1, \quad \frac{\partial w}{\partial r} = 0 \tag{3.106}$$

Integral Momentum Boundary-Layer Equations

The equations of motion (3.96) and (3.97) can be integrated within the boundary layer, using the equation of continuity and the other simplifications developed above to give the following:

$$\tau_{\mathrm{r\theta}}\big|_{\bar{\xi}=0} = -\frac{\rho}{a}\frac{d}{d\theta}\int_0^{\delta} v^2 d\bar{\xi} + 2\rho\Omega\sin\theta \int_0^{\delta} (w_1 - w) d\bar{\xi} \tag{3.107}$$

$$\tau_{\mathrm{rz}}\big|_{\bar{\xi}=0} = \frac{\rho}{a} w_1 \frac{d}{d\theta}\int_0^{\delta} v d\bar{\xi} - \frac{d}{d\theta}\int_0^{\delta} vw d\bar{\xi} \tag{3.108}$$

where $\bar{\xi} = a - r$ and consequently $\bar{\xi} = 0$ and $\bar{\xi} = \delta$ correspond to the pipe wall and the edge of the boundary layer, respectively.

Solution of the Momentum Integral Equations

In line with the tradition of the integral technique, suitable forms for the axial and angular velocity distributions in the boundary layer are to be chosen to obtain solutions of the momentum integral equations. This is done by first obtaining expressions for the local shear stress and the mean axial velocity distribution in the turbulent boundary-layer flow over a flat plate of zero angle of incidence and then extending it for the turbulent boundary-layer flow at hand.

The Blasius-type correlation developed (Shenoy & Mashelkar, 1983) for drag reducing fluids gives the following form for friction factor:

$$f = \frac{\alpha}{\mathrm{Re}^{\beta}} \tag{3.109}$$

In view of the above implication on the local velocity, the axial distribution in the boundary layer can be taken to be of the following form:

$$w = w_1 \left(\frac{\xi}{\delta}\right)^{\frac{\beta}{2-\beta}} \tag{3.110}$$

This equation can be seen to satisfy the boundary conditions set up in Eq. (3.106).

The angular velocity v changes from zero at the wall to a positive value and falls again to zero at the edge of the boundary layer. If near the wall v raises to D_1 (where D_1 is to be determined later) and at $\bar{\xi} = \delta$ it vanishes, we ought to choose the following form:

$$v = D_1 \left(\frac{\xi}{\delta}\right)^{\frac{\beta}{2-\beta}} \left(1 - \frac{\xi}{\delta}\right) \tag{3.111}$$

Resultant velocity in the neighborhood of the wall surface ($\bar{\xi} \geq 0$) can be found as

$$\lim_{\xi \to 0} \sqrt{v^2 + w^2} = U_1 \left(\frac{\xi}{\delta}\right)^{\frac{\beta}{2-\beta}} \tag{3.112}$$

where

$$U_1 = \sqrt{D_1^2 + w_1^2} \tag{3.113}$$

Using U_1 on the velocity scale very near the pipe wall, it is assumed that the semiempirical wall shear stress expression (Dodge & Metzner, 1959) remains valid if expressed in terms of U_1, for $n = 1$ and α and β given by the predicted values for drag reducing fluids (Shenoy & Mashelkar, 1983). Thus, similar to Eq. (3.40), the following can be written:

$$\tau_{\mathrm{w}} = \frac{\alpha[0.817]^{2-\beta}}{2^{1+\beta}} \rho U_1^{2-\beta} \left[\frac{\mu}{\rho\delta}\right]^{\beta} \tag{3.114}$$

Noting that

$$\frac{(\tau_{\mathrm{r}\theta})_{\xi=0}}{(\tau_{\mathrm{r}\varnothing})_{\xi=0}} = \lim_{\xi\to 0} \frac{u}{w} = \frac{D_1}{w_1} \tag{3.115}$$

and

$$\tau_{\mathrm{w}} = \sqrt{(\tau_{\mathrm{r}\theta})_{\xi=0}^2 + (\tau_{\mathrm{rz}})_{\xi=0}^2} \tag{3.116}$$

gives

$$\tau_{\mathrm{r}\theta} = \frac{D_1}{\sqrt{D_1^2 + w_1^2}} \tau_{\mathrm{w}} \tag{3.117}$$

and

$$\tau_{\mathrm{rz}} = \frac{w_1}{\sqrt{D_1^2 + w_1^2}} \tau_{\mathrm{w}} \tag{3.118}$$

An expression for w_1 can be obtained from Eq. (3.86) using the thin boundary-layer assumption as

$$w_1 = w_{\mathrm{m}} + \overline{B}a \cos\theta \tag{3.119}$$

Using the following forms for nondimensionalizing

$$\frac{\delta}{a} = \delta_{\mathrm{r}} \left[\mathrm{Re}\left(\frac{2a\overline{\Omega}}{w_{\mathrm{m}}}\right)^{\frac{1}{2\beta}}\right]^{\frac{-\beta}{\beta+1}} \tag{3.120}$$

$$\frac{D_1}{D_{\mathrm{r}}} = \left(2a\overline{\Omega}w_{\mathrm{m}}\right)^{\frac{1}{2}} \tag{3.121}$$

$$\frac{w_1}{w_{\mathrm{r}}} = w_{\mathrm{m}} \tag{3.122}$$

Equations (3.107) and (3.108) reduce to the following:

$$f_1 \frac{\partial}{\partial \theta}\left[D_r^2 \delta_r\right] = [1 - f_2]\delta_r w_r \sin\theta - f_3 \delta_r^{-\beta} D_r w_r^{1-\beta} \tag{3.123}$$

$$f_4 \frac{\partial}{\partial \theta}[D_r w_r \delta_r] - f_5 w_r \frac{\partial}{\partial \theta}[D_r \delta_r] = -f_3 \delta_r^{-\beta} w_r^{1-\beta} \tag{3.124}$$

where

$$f_1 = \frac{(2-\beta)^3}{2(2+\beta)(6-\beta)} \tag{3.125}$$

$$f_2 = \frac{(2-\beta)}{(2+\beta)} \tag{3.126}$$

$$f_3 = \frac{\alpha(0.817)^{2-\beta}}{2^{\beta+1}} \tag{3.127}$$

$$f_4 = \frac{(2-\beta)^2}{2(4-\beta)} \tag{3.128}$$

$$f_5 = \frac{(2-\beta)^2}{4(2+\beta)} \tag{3.129}$$

Since α and β are functions of De for dilute drag reducing fluids, f_1 to f_5 are all functions of De alone as reported in Table 3.4.

The nondimensional form of Eq. (3.119) can be written as

$$w_r = 1 + \frac{\overline{B}a}{w_m} \cos\theta \tag{3.130}$$

where from Eq. (3.91)

$$\overline{B} = \frac{\alpha^*}{\pi f_5} \int_0^{\pi} \frac{\sin\theta}{D_r \delta_r} d\theta \tag{3.131}$$

and

$$\alpha^* = \frac{2f_3}{\pi} \int_0^{\pi} w_r^{2-\beta} \delta_r^{-\beta} d\theta \tag{3.132}$$

The relationship for friction factor is now obtained by using Eq. (3.94) along with the definition of average friction factor:

$$f_r = \frac{4a\left(-\frac{\partial p^*}{\partial z}\right)}{\rho w_m^2} \tag{3.133}$$

Table 3.4 Values of various parameters for flow of dilute drag reducing fluids through a tube rotating about an axis perpendicular to its own (Shenoy, 1986)

De	α	β	f_1	f_2	f_3	f_4	f_5	δ_r	$\alpha*$	$\frac{\overline{B}\,a}{w_m}$
0	0.0790	0.25	0.2071	0.8750	0.02773	0.4083	0.3403	0.5623	0.0617	0.4271
2	0.0767	0.271	0.1986	0.8645	0.02704	0.4008	0.3291	0.4694	0.0497	0.4486
4	0.0726	0.285	0.1931	0.8575	0.02567	0.3959	0.3218	0.3920	0.0424	0.4693
6	0.0655	0.292	0.1904	0.8540	0.02319	0.3934	0.3182	0.3508	0.0392	0.4900
10	0.0762	0.334	0.1748	0.8330	0.02721	0.3786	0.2973	0.3162	0.0404	0.5317

which upon substitution of the expression for $\frac{\partial p^*}{\partial z}$ from Eq. (3.94) gives

$$f_r = \frac{4\alpha^* \left(\frac{2a\overline{\Omega}}{w_m}\right)^{\frac{1}{2}}}{\left[\mathrm{Re}\left(\frac{2a\overline{\Omega}}{w_m}\right)^{\frac{1}{2\beta}}\right]^{\frac{\beta}{\beta+1}}} \tag{3.134}$$

Thus, it can be seen that the appropriate similarity variable in the case of drag reducing fluids turns out to be $\mathrm{Re}\left(\frac{2a\overline{\Omega}}{w_m}\right)^{\frac{1}{2\beta}}$ where β is a function of De.

Method of Solution

Equations (3.123), (3.124), (3.125), (3.126), (3.127), (3.128), (3.129), (3.130), (3.131), and (3.132) form a set of complex equations which cannot be solved analytically, and numerical techniques must be resorted to. Initial values of δ_r, D_r, and w_r needed to start the integration are obtained by expanding in the neighborhood of $\theta = 0$ as follows:

$$\delta_r = \delta_1\left(1 + \delta_2\theta^2 + \cdots\right) \tag{3.135}$$

$$D_r = D_1\theta\left(1 + D_3\theta^3 + \cdots\right) \tag{3.136}$$

$$w_r = 1 + \frac{\overline{B}\,a}{w_m}\left(1 - \frac{\theta^2}{2!} + \frac{\theta^4}{4!} - \cdots\right) \tag{3.137}$$

Substituting these into Eqs. (3.123) and (3.124) and rearranging and equating equal powers of δ yield

$$\delta_1{}^{-2\beta} = 2[1 - f_2]\left[1 + \frac{\overline{B}a}{w_m}\right]^{\beta-1}\left[\frac{2f_1 f_3^2}{(f_4 - f_5)^2} - \frac{f_4^2}{f_4 - f_5}\right] \tag{3.138}$$

$$D_1 = \left[-\frac{f_3}{f_4 - f_5}\right] w_r^{1-\beta}\delta_r^{-\beta} \tag{3.139}$$

To start the integration procedure, the value of $\frac{\overline{B}a}{w_m}$ is first assumed so that δ_1 is known. The numerical integration is then started for obtaining $\delta_0(\theta)$ and $D_0(\theta)$ from Eqs. (3.123) and (3.124) using a Runge-Kutta-Merson technique. Substitution of these $\delta_0(\theta)$ and $D_0(\theta)$ into Eq. (3.132) gave a value of $\alpha*$ which is in turn used to calculate $\frac{\overline{B}a}{w_m}$ from Eq. (3.131). The obtained value is compared with the initial assumed value, and the computations are continued until the difference between the two values is less than 10^{-4}. The calculations are performed for Deborah numbers De = 0, 2, 4, 6, 10.

Discussion of the Numerical Results

Table 3.4 gives the values of the functions α, β, and f_1 to f_5 and the computed values of δ_1, $\alpha*$, and $\frac{\overline{B}a}{w_m}$ for dilute drag reducing fluids ($0 \leq \text{De} \leq 10$). Excellent agreement is found on comparing the computed values of δ_1, $\alpha*$, and $\frac{\overline{B}a}{w_m}$ for Newtonian fluids (De = 0) (Nanbu, 1970) giving evidence of the propriety of the numerical scheme.

The numerical results of δ_r, w_r, and D_r are plotted as functions of θ in Figs. 3.8, 3.9, and 3.10, for fluids with different Deborah numbers.

Figure 3.8 shows that there is a marked decreasing trend in the dimensionless boundary-layer thickness with increasing elasticity. However, w_0 and D_0 are affected to a much lesser extent by elasticity as can be seen from Figs. 3.9 and 3.10.

Figures 3.8, 3.9, and 3.10 along with Eqs. (3.120), (3.121), (3.122), and (3.86) provide the complete velocity distribution for turbulent flow of dilute drag reducing fluids through a tube rotating around an axis perpendicular to its own.

It is worth realizing the pragmatic importance of Eq. (3.134) as it provides a very easy means of predicting frictional losses in a tube rotating around an axis perpendicular to its own knowing the values of α, β, and $\alpha*$ from Table 3.4 for dilute drag reducing fluids.

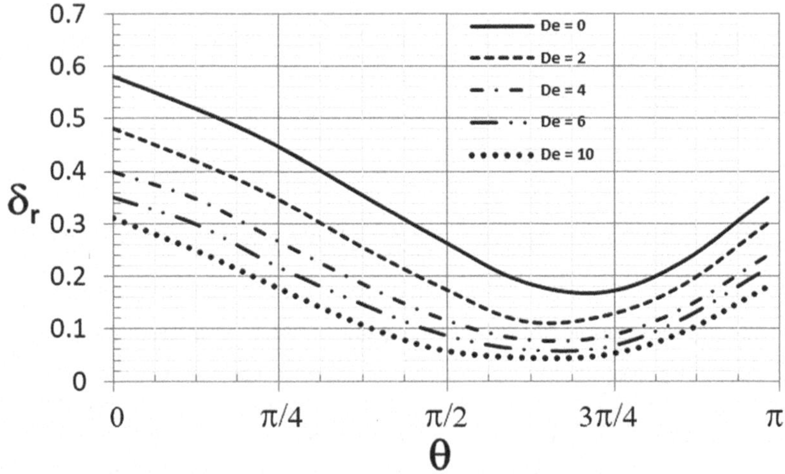

Fig. 3.8 Variation of the boundary-layer thickness δ_r with θ and De (Shenoy, 1986). (Reprinted with permission from Springer and Copyright Clearance Center through RightsLink)

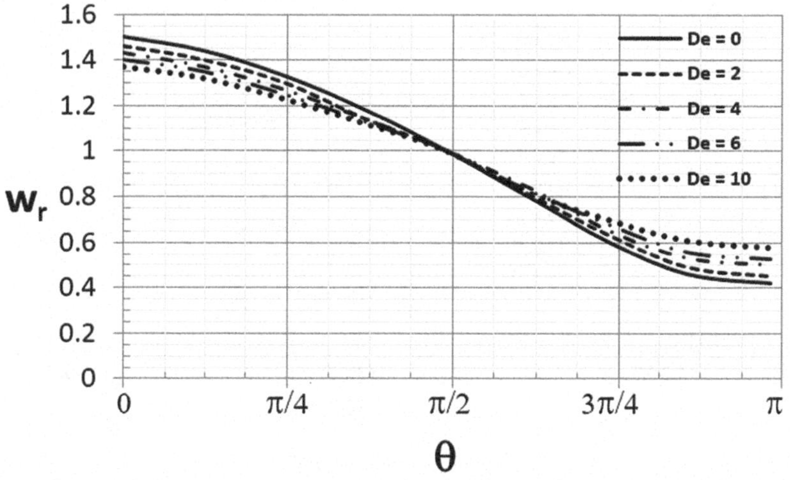

Fig. 3.9 Variation of the axial velocity w_r with θ and De (Shenoy, 1986). (Reprinted with permission from Springer and Copyright Clearance Center through RightsLink)

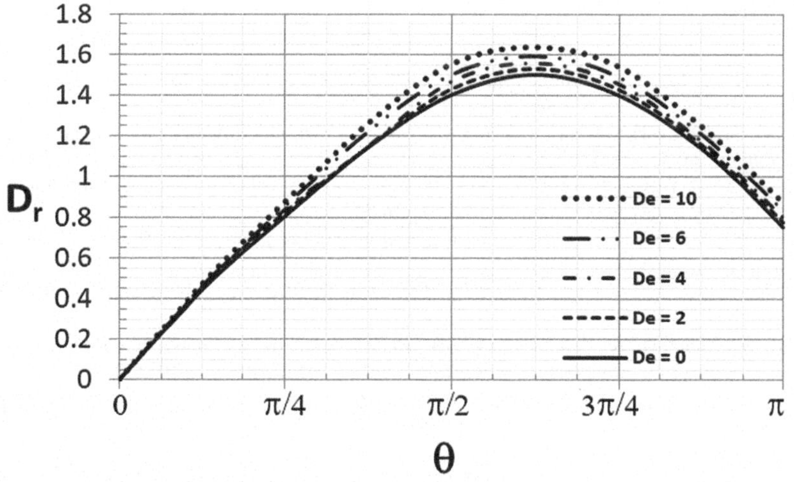

Fig. 3.10 Variation of the angular velocity D_r with θ and De (Shenoy, 1986). (Reprinted with permission from Springer and Copyright Clearance Center through RightsLink)

Figure 3.11 shows the effect of Deborah number De on the friction factor versus Reynolds number with change in the angular velocity. The increasing value of De indicates increasing fluid elasticity because the relaxation time θ_{fl} appearing in the definition of De represents the time required for the polymer molecules to relax. Thus, higher elasticity results in lower values of friction factor which is basically the well-known effects of the phenomenon of drag reduction.

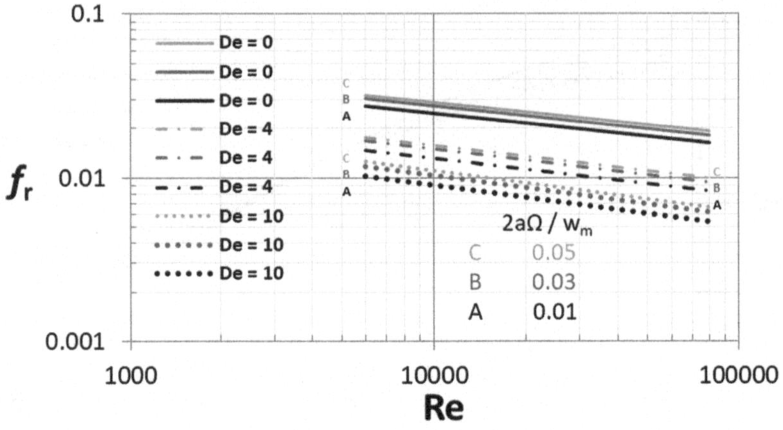

Fig. 3.11 Variation of the friction factor f_r with Reynolds number Re for different angular velocities and De (Shenoy, 1986). (Reprinted with permission from Springer and Copyright Clearance Center through RightsLink)

References

Barua, S. H. (1963). On secondary flow in stationary curved pipes. *The Quarterly Journal of Mechanics and Applied Mathematics, 14*, 61–77.

Benton, G. S. (1956). The effects of the earth's rotation on laminar flow in pipes. *Journal of Applied Mechanics, 23*, 123–127.

Benton, G. S., & Boyer, D. (1966). Flow through a rapidly rotating conduit of arbitrary cross-section. *Journal of Fluid Mechanics, 26*(1), 69–79.

Devarajan, G. V. (1976). *Secondary flows of non-newtonian fluids* (PhD Thesis). University of Salford, U.K.

Dodge, D. W., & Metzner, A. B. (1959). Turbulent flow of non-newtonian systems. *AICHE Journal, 5*(2), 189–204.

Gunn, R. W., Mena, B., & Walters, K. (1974). On newtonian and non-newtonian flow in a rotating pipe. *Zeitschrift fur Angewandte Mathematik und Physik (ZAMP), 25*(5), 591–606.

Ito, H. M. (1956). *Reports of the institute of high speed mechanics.* Tohoku University, Japan.

Ito, H. M. (1959a). Memoirs of the Institute of High Speed Mechanics. *Tohoku University, Japan, 14*, 137.

Ito, H. M. (1959b). Friction factors for turbulent flow in curved pipes. *Transactions ASME, Journal of Basic Engineering, 81*(2), 123–132.

Ito, H. M., & Nanbu, K. (1971). Flow in rotating straight pipes of circular cross section. *Transactions ASME Journal of Basic Engineering, 93*(3), 383–394.

Jones, D. T., Trevors, J. R., & Walters, K. (1967). A note on the motion of a viscous liquid in a rotating straight pipe. *ZAMP, 18*(6), 774–781.

Mori, I., & Nakayama, W. (1968). Convective heat transfer in rotating radial circular pipes. *International Journal of Heat and Mass Transfer, 11*(6), 1027–1040.

Nanbu, K. (1970). *Memoirs of institute of high speed mechanics. Tohoku University, Japan, 25* (262).

Oliver, D. R. (1966). The expansion/contraction behavior of laminar liquid jets. *The Canadian Journal of Chemical Engineering, 44*(2), 100–107.

Seyer, F. A., & Metzner, A. B. (1969). Drag reduction in large tubes and the behavior of annular films of drag reducing fluids. *The Canadian Journal of Chemical Engineering, 47*(6), 525–529.

Shenoy, A. V. (1986). Turbulent flow of mildly elastic fluids through rotating straight circular tubes. *Journal of Applied Sciences Research, 43*(1), 39–54.

Shenoy, A. V., & Mashelkar, R. A. (1983). Engineering estimate of hydrodynamic entrance lengths in non-newtonian turbulent flow. *Industrial and Engineering Chemistry Process Design and Development, 22*(1), 165–168.

Shenoy, A. V., Ranade, V. R., & Ulbrecht, J. J. (1980). Turbulent flow of mildly viscoelastic liquids in curved tubes. *Chemical Engineering Communications, 5*(5–6), 269–286.

Chapter 4
Turbulent Natural Convection Heat Transfer in External Flows

Vertical Flat Plate

The theoretical analysis of turbulent natural convection for mildly elastic drag reducing fluids is done in a manner similar to that for Newtonian fluids (Eckert & Jackson, 1950). The approximate integral method is used to obtain the asymptotic solution for high Prandtl numbers using a similarity transformation.

The plate is assumed to be vertical, flat, and semi-infinite as shown in Fig. 4.1, and the physical properties of the fluid (except the density in the buoyancy term) are assumed to be constant.

The integral equations for momentum and heat balances in the boundary layer can be set down in a manner similar to those for Newtonian fluids (Eckert & Jackson, 1950) as follows:

$$\frac{d}{dx}\int_0^\delta u^2 dy = g\beta_0 \int_0^\delta (T - T_\infty)dy - \frac{\tau_w}{\rho} \tag{4.1}$$

$$\frac{d}{dx}\int_0^\delta uTdy = -\frac{k}{\rho C_p}\left(\frac{\partial T}{\partial y}\right)_w \tag{4.2}$$

The boundary conditions on the velocity and temperature are as follows:

$$u(x,0) = u(x,\delta) = 0; \quad T(x,0) = T_w; \quad T(x,\delta) = T_\infty \tag{4.3}$$

The Blasius-type correlation developed (Shenoy & Mashelkar, 1983) for drag reducing fluids (Eq. (2.25)) gives the following form for friction factor in smooth pipe flow:

$$f = \frac{\alpha}{\mathrm{Re}^\beta} \text{ for } 5\times 10^3 \le \mathrm{Re} \le 10^5 \text{ and } 0 \le \mathrm{De} \le 10 \tag{4.4}$$

A. Shenoy, *Rheology of Drag Reducing Fluids*,
https://doi.org/10.1007/978-3-030-40045-3_4

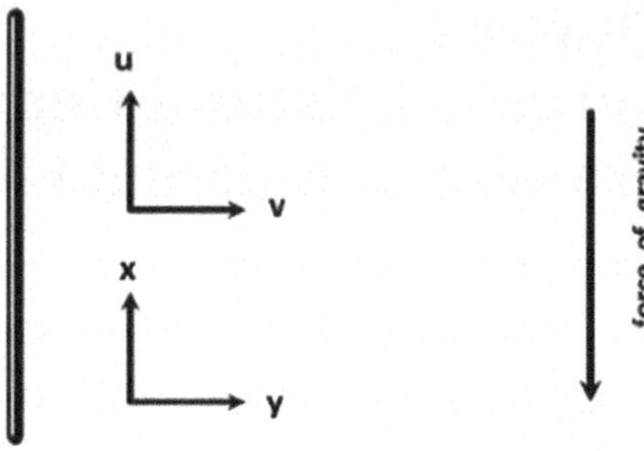

Fig. 4.1 Schematic diagram of flow past a vertical flat plate

α and β are both functions of Deborah number, and their values for varying De are given (Shenoy & Shintre, 1986) and shown in Table 2.1.

A suitable expression for the local surface shear stress can be obtained by proper rearranging and adapting the equations to flow over a smooth flat plate at zero incidence in a manner analogous to that used in the Newtonian case (Eckert & Jackson, 1950) as

$$\tau_w = \Omega \rho^{1-\beta} \mu^{\beta} \delta^{-\beta} \Lambda_1^{2-\beta} \tag{4.5}$$

where

$$\Omega = \frac{\alpha (0.817)^{2-\beta}}{2^{\beta+1}} \tag{4.6}$$

The Colburn's analogy between heat and momentum transfer may now be applied to the turbulent flow over the flat plate as

$$\frac{h_w}{C_p \Lambda_1} \left(\frac{C_p \mu}{k} \right)^{\frac{2}{3}} = \frac{\tau_w}{\rho \Lambda_1^2} \tag{4.7}$$

Thus yielding

$$h_w = \Omega C_p \rho^{1-\beta} \delta^{-\beta} \Lambda_1^{1-\beta} \left(\frac{C_p \mu}{k} \right)^{-\frac{2}{3}} \tag{4.8}$$

With the assumption made for Newtonian fluids (Eckert & Jackson, 1950) that close to the wall, the relationships connecting wall shear stress and heat flow with temperatures and velocities in this range are the same for forced flow and natural convection flow, the following can be used:

$$-k \left(\frac{\partial T}{\partial y} \right)_w = q_w = h_w (T_w - T_\infty) \tag{4.9}$$

Using Eqs. (4.5), (4.8), and (4.9), the simplified forms of Eqs. (4.1) and (4.2) can now be obtained as follows:

$$\frac{d}{dx}\int_0^{\delta} u^2 dy = g\beta_0 \int_0^{\delta} (T - T_\infty)dy - \Omega\rho^{1-\beta}\mu^{\beta}\delta^{-\beta}\Lambda_1^{2-\beta} \tag{4.10}$$

$$\frac{d}{dx}\int_0^{\delta} uT dy = (T_w - T_\infty)\Omega\rho^{-\beta}\mu^{\beta}\delta^{-\beta}\Lambda_1^{1-\beta}\left(\frac{C_p\mu}{k}\right)^{-\frac{2}{3}} \tag{4.11}$$

An order of magnitude analysis of Eqs. (4.10) and (4.11) can be carried out using $u \sim O(U_c)$, $x \sim O(l_c)$, $y \sim O(\delta)$, and $\Lambda_1 \sim O(U_c)$. It is easily seen that for large values of a characteristic Prandtl number Pr_c defined as

$$\mathrm{Pr}_c = \frac{C_p}{k}\mu^{4\beta}\rho^{1-4\beta} l_c^{\frac{3(1-4\beta)}{2}} [g\beta_0(T_w - T_\infty)]^{\frac{1-4\beta}{2}} \tag{4.12}$$

for the constant temperature plate, the inertial terms in the momentum equation are negligible in comparison to the other terms on the right-hand side of Eq. (4.10).

A characteristic Grashof number is now defined by taking the ratio of the buoyancy force to the viscous force as

$$\mathrm{Gr}_c = \mu^{-8\beta}\rho^{8\beta} l_c^{12\beta} [g\beta_0(T_w - T_\infty)]^{4\beta} \tag{4.13}$$

As there exists no characteristic length for the external flow past the semi-infinite plate under consideration, l_c is chosen such that $\mathrm{Gr}_c = 1$ (Hellums & Churchill, 1964). Thus,

$$l_c = \left\{\frac{\mu^2}{\rho^2[g\beta_0(T_w - T_\infty)]}\right\}^{\frac{1}{3}} \tag{4.14}$$

$$U_c = \left\{\frac{\mu[g\beta_0(T_w - T_\infty)]}{\rho}\right\}^{\frac{1}{3}} \tag{4.15}$$

The nondimensional variables are now defined as

$$x_1 = \frac{x}{l_c}, \quad y_1 = \frac{y}{l_c}, \quad \delta_1 = \frac{\delta}{l_c} \tag{4.16}$$
$$u_1 = \frac{u}{U_c}, \quad A_1 = \frac{\Lambda_1}{U_c}, \quad \theta_1 = \frac{T - T_\infty}{T_w - T_\infty}$$

The nondimensional forms of Eq. (4.10) (on neglecting inertia) and Eq. (4.11) can now be written as

$$0 = \int_0^{\delta_1} \theta_1 dy - \Omega \delta_1^{-\beta} A_1^{2-\beta} \tag{4.17}$$

$$\frac{d}{dx_1} \int_0^{\delta_1} u_1 \theta_1 dy_1 = \Omega \left(\frac{0.02332}{\Omega} \right)^{\frac{8}{3}} \mathrm{Gr}_x^{\frac{1-4\beta}{12\beta}} \mathrm{Pr}_x^{-\frac{2}{3}} \delta_1^{\frac{5\beta-2}{3}} A_1^{1-\beta} \tag{4.18}$$

where Gr_x is the local distance-based Grashof number defined as

$$\mathrm{Gr}_x = \mu^{-8\beta} \rho^{8\beta} x^{12\beta} [g\beta_0 (T_w - T_\infty)]^{4\beta} \tag{4.19}$$

and Pr_x is the local distance-based Prandtl number defined as

$$\mathrm{Pr}_x = \frac{C_p}{k} \mu^{4\beta} \rho^{1-4\beta} x^{\frac{3(1-4\beta)}{2}} [g\beta_0 (T_w - T_\infty)]^{\frac{1-4\beta}{2}} \tag{4.20}$$

Note that $\mathrm{Gr}_x^{\frac{1-4\beta}{12\beta}} \mathrm{Pr}_x^{-\frac{2}{3}}$ is independent of x and will be treated like a constant during the following analysis.

Equations (4.17) and (4.18) are now solved for the following boundary conditions:

$$\begin{gathered} u_1(x_1, 0) = u_1(x_1, \delta_1) = 0 \\ \theta_1(x_1, 0) = 1; \quad \theta_1(x_1, \delta_1) = 0 \end{gathered} \tag{4.21}$$

Expressions for u_1 and θ_1 are sought in a manner similar to that for Newtonian fluids (Eckert & Jackson, 1950). They noted that the equations $u = \Lambda\left(\frac{y}{\delta}\right)^{\frac{1}{7}}$ and $(T - T_\infty) = (T_w - T_\infty)\left\{1 - \left(\frac{y}{\delta}\right)^{\frac{1}{7}}\right\}$ hold considerably well in forced convection. Observing temperature and velocity distributions obtained experimentally (Griffiths & Davis, 1922), they concluded that the temperature equation fitted the free convection experimental data reasonably well, while, of course, the velocity profile showed a different trend because in free convection the velocity is zero both at the solid surface and remote from it. However, they found that the equation $u = \Lambda\left(\frac{y}{\delta}\right)^{\frac{1}{7}}\left\{1 - \left(\frac{y}{\delta}\right)\right\}^4$ fitted the shape quite well.

For mildly elastic drag reducing fluids, the velocity profile for turbulent forced convection can be assumed as $u = \Lambda(y/\delta)^q$, where $q = \beta/[2 - \beta]$ along the lines for power-law fluids (Skelland, 1967). In the free convection case, the velocity and temperature profiles are assumed by analogous arguments to those for Newtonian fluids (Eckert & Jackson, 1950), making use of the forced convection expression for power-law fluids (Skelland, 1967). Thus, the dimensionless temperature and velocity profiles which are assumed to fit the turbulent free convection flow of mildly elastic drag reducing fluids are

$$\theta_1(\eta) = 1 - \eta^q \tag{4.22}$$

$$u_1(\eta) = A\eta^q\{1-\eta\}^4 \tag{4.23}$$

where

$$\eta = \frac{y_1}{\delta_1} \tag{4.24}$$

and

$$q = \frac{\beta}{2-\beta} \tag{4.25}$$

Substituting Eqs. (4.22) and (4.23) in Eqs. (4.17) and (4.18), and appropriately rearranging the terms, gives

$$0 = C_1\delta_1 - \Omega\delta_1^{-\beta}A_1^{2-\beta} \tag{4.26}$$

$$C_2\frac{d}{dx_1}(\delta_1 A_1) = \Omega\left(\frac{0.02332}{\Omega}\right)^{\frac{8}{3}}\mathrm{Gr}_x^{\frac{1-4\beta}{12\beta}}\mathrm{Pr}_x^{-\frac{2}{3}}\delta_1^{\frac{5\beta-2}{3}}A_1^{\frac{5\beta+1}{3}} \tag{4.27}$$

where

$$C_1 = \frac{q}{q+1} \tag{4.28}$$

$$C_2 = \frac{3}{q+1} - \frac{2}{q+2} + \frac{6}{q+3} - \frac{4}{q+4} + \frac{1}{q+5} - \frac{1}{2q+1} - \frac{6}{2q+3} - \frac{1}{2q+5} \tag{4.29}$$

For a similarity search, the following forms of δ_1 and A_1 are assumed:

$$\delta_1 = B_1 x_1^{\bar{r}} \tag{4.30}$$

$$A_1 = B_2 x_1^{\bar{s}} \tag{4.31}$$

Substituting these into Eqs. (4.26) and (4.27) and equating the powers of x_1 for the equations to be valid for any x_1 give

$$\bar{r} = \frac{(2-\beta)}{2(2-3\beta)} \tag{4.32}$$

$$\bar{s} = \frac{(1+\beta)}{2(2-3\beta)} \tag{4.33}$$

thus, giving the conditions for similarity. Note that $\bar{r}$ and $\bar{s}$ reduce correctly to their respective values of 7/10 and 1/2 when $\beta = 0.25$ as obtained for Newtonian fluids (Eckert & Jackson, 1950).

The simplified forms of Eqs. (4.26) and (4.27) on substitution of Eqs. (4.30), (4.31), (4.32), and (4.33) are

$$C_1 B_1^{1+\beta} = \Omega B_2^{2-\beta} \tag{4.34}$$

$$B_1^{\frac{5(1-\beta)}{3}} B_2^{\frac{2-5\beta}{3}} = \frac{\Omega}{C_2}\left(\frac{0.02332}{\Omega}\right)^{\frac{8}{3}}\left\{\frac{2(2-3\beta)}{3}\right\}\mathrm{Gr}_{\mathrm{x}}^{\frac{1-4\beta}{12\beta}}\mathrm{Pr}_{\mathrm{x}}^{-\frac{2}{3}} \tag{4.35}$$

Solving (4.34) and (4.35) gives

$$\begin{aligned} B_1^{\frac{5(1-\beta)}{3}} B_2^{\frac{2-5\beta}{3}} &= \left(\frac{\Omega}{C_1}\right)^{\frac{2-5\beta}{6(2-3\beta)}}\left(\frac{\Omega}{C_2}\right)^{\frac{2-\beta}{2(2-3\beta)}} \\ &\left(\frac{0.02332}{\Omega}\right)^{\frac{4(2-\beta)}{3(2-3\beta)}}\left\{\frac{2(2-3\beta)}{3}\right\}^{\frac{2-\beta}{2(2-3\beta)}}\mathrm{Gr}_{\mathrm{x}}^{\frac{(1-4\beta)(2-\beta)}{24\beta(2-3\beta)}}\mathrm{Pr}_{\mathrm{x}}^{-\frac{2-\beta}{2(2-3\beta)}} \end{aligned} \tag{4.36}$$

$$B_2 = \left(\frac{C_1}{\Omega}\right)^{\frac{1}{2-\beta}}(B_1)^{\frac{1+\beta}{2-\beta}} \tag{4.37}$$

The local Nusselt number is defined as

$$\mathrm{Nu}_{\mathrm{x}} = \frac{h_{\mathrm{w}}x}{k} = \Omega\left(\frac{0.02332}{\Omega}\right)^{\frac{8}{3}}(\delta_1)^{\frac{5\beta-2}{3}}(A_1)^{\frac{1+5\beta}{3}}\mathrm{Gr}_{\mathrm{x}}^{\frac{1+4\beta}{24\beta}}\mathrm{Pr}_{\mathrm{x}}^{\frac{1}{3}} \tag{4.38}$$

Substituting appropriately and simplifying give

$$\begin{aligned} \mathrm{Nu}_{\mathrm{x}} &= \frac{h_{\mathrm{w}}x}{k} = (0.02332)^{\frac{4}{2-3\beta}}\Omega^{-\frac{3}{2-3\beta}}(C_1)^{\frac{1}{2(2-3\beta)}}(C_2)^{\frac{1-6\beta}{2(2-3\beta)}} \\ &\left\{\frac{2(2-3\beta)}{3}\right\}^{-\frac{1-6\beta}{2(2-3\beta)}}\mathrm{Gr}_{\mathrm{x}}^{\frac{7-12\beta}{8(2-3\beta)}}\mathrm{Pr}_{\mathrm{x}}^{\frac{1-3\beta}{2-3\beta}} \end{aligned} \tag{4.39}$$

Equation (4.39) can be expressed as

$$\mathrm{Nu}_{\mathrm{x}} = C\mathrm{Gr}_{\mathrm{x}}^{a'}\mathrm{Pr}_{\mathrm{x}}^{b'} \tag{4.40}$$

with the appropriate definitions of C, a', and b', the values of which for varying β are tabulated in Table 4.1. Some of the columns in Table 4.1 have been given to five significant digits in order to maintain consistency with values customarily well-known for Newtonian fluids.

With increasing Deborah number, the coefficient C increases till De $= 6$ and decreases, while the exponents a' and b' decrease continuously. But the trend of

Table 4.1 Values of various parameters for varying values of Deborah numbers for turbulent free convection from a vertical flat plate

De	α	β	Ω	q	$C1$	$C2$	C	a'	b'
0	0.079	0.25	0.02332	0.14286	0.1250	0.0366	0.0402	0.400	0.200
1	0.0782	0.26	0.02295	0.15075	0.1310	0.0375	0.0420	0.397	0.176
2	0.0787	0.27	0.02299	0.15674	0.1355	0.0382	0.0419	0.395	0.158
3	0.0741	0.28	0.02157	0.16144	0.1390	0.0386	0.0493	0.393	0.142
4	0.0726	0.29	0.02107	0.16618	0.1425	0.0391	0.0526	0.391	0.127
5	0.0689	0.29	0.01995	0.16891	0.1445	0.0393	0.0608	0.390	0.117
6	0.0655	0.29	0.01894	0.17096	0.1460	0.0395	0.0699	0.389	0.110
7	0.0662	0.3	0.01906	0.17716	0.1505	0.0400	0.0692	0.386	0.088
8	0.0687	0.31	0.01967	0.18483	0.1560	0.0406	0.0637	0.383	0.060
9	0.0732	0.32	0.02084	0.19332	0.1620	0.0412	0.0538	0.378	0.027
10	0.0762	0.33	0.02158	0.20048	0.1670	0.0417	0.0480	0.375	−0.002

Nusselt number cannot be easily ascertained owing to the unknown Grashof and Prandtl number changes with increasing Deborah number. Nevertheless, Eq. (4.40) would be very useful in the design of turbulent natural convection processes in mildly elastic drag reducing fluids. With the assumption that the boundary layer is turbulent over the whole of the plate, the average Nusselt number can be easily written by taking an integrated average over the length of the plate L as follows:

$$\mathrm{Nu}_{\mathrm{av}} = \frac{h_{\mathrm{w}}L}{k} = (0.02332)^{\frac{4}{2-3\beta}}\Omega^{-\frac{3}{2-3\beta}}(C_1)^{\frac{1}{2(2-3\beta)}}(C_2)^{\frac{1-6\beta}{2(2-3\beta)}} \left\{\frac{2(2-3\beta)}{3}\right\}^{\frac{3}{2(2-3\beta)}}\mathrm{Gr}_{\mathrm{L}}^{\frac{7-12\beta}{8(2-3\beta)}}\mathrm{Pr}_{\mathrm{L}}^{\frac{1-3\beta}{2-3\beta}} \tag{4.41}$$

In reality, of course, the boundary layer is initially laminar and becomes turbulent only at a certain distance from the leading edge of the plate. Equation (4.41) could be expected to predict true average Nusselt number values only at Grashof numbers which are large enough such that the extent of the laminar boundary layer at the lower edge of the plate is small compared with the total length L of the plate. This limit for the Grashof number has been suggested to be around 10^{10} for Newtonian fluids (Eckert & Jackson, 1950) and could be assumed to be around the same for the case of mildly elastic drag reducing fluids.

Finally, it is worth mentioning that although the development in the present analysis is closely based on the Newtonian fluids counterpart (Eckert & Jackson, 1950), Eqs. (4.40) and (4.41) differ in structure from the expressions for the Newtonian case owing to the extra assumption of high Prandtl number made herein and hence could be matched only when a similar assumption is made in the final forms of the local and average Nusselt numbers for Newtonian fluids as given in Eq. (4.91).

Arbitrary Geometric Configurations

The turbulent free convection problem from bodies of arbitrary geometric configurations to mildly elastic drag reducing fluids (Nakayama & Shenoy, 1992) is presented below.

It is assumed that the geometric configuration has an arbitrary shape and the coordinate system is as shown in Fig. 4.2.

The body may be planar or axisymmetric, and its wall geometry is defined by the function $r(x)$. The wall surface is heated to $T_w(x)$ above the ambient temperature T_∞, which is assumed to be constant. The flow is induced against the gravitational force g under the influence of the buoyancy force component parallel to the wall surface. The appearance of turbulence in the flow begins at the top of the surface and gradually extends to cover more and more of the surface as the Grashof number increases. Turbulence occurs when the surface in question is big or the temperature difference is large.

A usual control volume analysis within the boundary layer of thickness δ leads to the following integral forms of the momentum and energy equations under the Boussinesq approximation on the buoyancy force:

$$\frac{d}{dx}\int_0^\delta r^*\rho u^2 dy = r^*\rho g_x \beta_0 \int_0^\delta (T - T_e)dy - r^*\tau_w \tag{4.42}$$

$$\frac{d}{dx}\int_0^\delta r^*\rho u C_p (T - T_e)dy = r^* q_w \tag{4.43}$$

where

$$r^* = 1 \text{ (for planar flow)} \tag{4.44a}$$

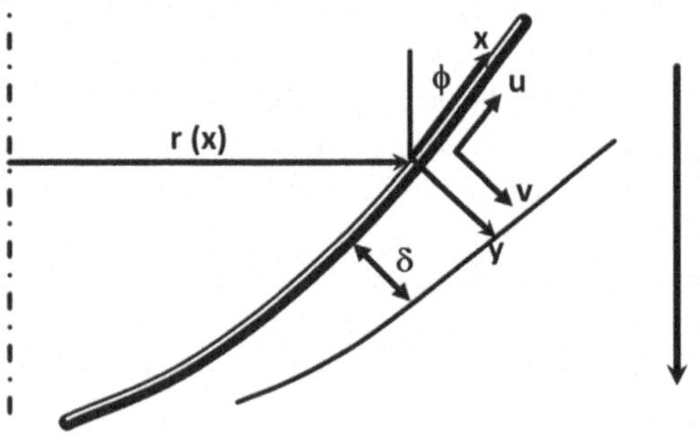

Fig. 4.2 Schematic diagram of flow past an arbitrary geometric configuration along the with the coordinates (Nakayama & Shenoy, 1992). (Reprinted with permission from ASME)

$$r^* = r(x) \text{ (for axisymmetric flow)} \tag{4.44b}$$

and

$$g_x = g \cos\varphi = g\left\{1 - \left(\frac{dr}{dx}\right)^2\right\}^{\frac{1}{2}} \tag{4.45}$$

In the above equations, τ_w and q_w are the local wall shear and heat flux, while ρ, C_p, and β_0 are the density, specific heat, and thermal expansion coefficient, respectively. The tangential component of the acceleration due to gravity is indicated by g_x which is related to the local surface orientation φ through Eq. (4.45). Moreover, the streamwise velocity and the local wall temperature are denoted by u and T with the subscripts ∞ and w specifically pertaining to the boundary-layer edge and beyond and the wall surface, respectively.

Theoretical analysis of free convection is normally more difficult than forced convection due to the coupling of the momentum and energy equation in the former case. When dealing with non-Newtonian fluids, this task becomes even more formidable. Hence, certain simplifications are sought in order to facilitate a solution without much sacrifice of accuracy. The first step is to see how the equations would scale if the flow was purely forced convection. When dealing with turbulent non-Newtonian boundary-layer flow, it is known that there exists a viscous sublayer that is very thin and close to the wall where the flow would be akin to a laminar forced convection flow. Also at the wall, it is obvious that the local shear stress and the local heat flux assume their maximum value. Thus, an order of magnitude analysis of the kind used for Newtonian fluids (Bejan, 1984) can be used for the present case:

$$\tau_w \sim \rho U_c^2 \left(\frac{\rho U_c l_c}{\mu}\right)^{-\frac{1}{2}} \tag{4.46}$$

$$\frac{q_w}{(T_w - T_e)} \sim \frac{k}{l_c} \mathrm{Pr}^{\frac{1}{3}} \left(\frac{\rho U_c l_c}{\mu}\right)^{\frac{1}{2}} \tag{4.47}$$

where U_c is the characteristic velocity and Pr is the Prandtl number for drag reducing fluids defined in the conventional manner as given below:

$$\mathrm{Pr} = \frac{\mu C_p}{k} \tag{4.48}$$

Combining Eqs. (4.46) and (4.47) gives the following:

$$\frac{q_w}{\rho C_p (T_w - T_e) U_c} \sim \frac{\tau_w}{\rho u_c^2} \mathrm{Pr}^{-\frac{2}{3}} \tag{4.49}$$

It is now assumed that the above equation would hold even for the free convection flow if the characteristic velocity is related to the buoyancy rather than the free-stream velocity as in the forced convection case.

The dimensionless functions u_1 and θ_1 for the velocity and temperature profiles are introduced as follows:

$$u_1(\eta) = \frac{u}{U_c} \tag{4.50}$$

$$\theta_1(\eta) = \frac{(T - T_e)}{\Delta T} \tag{4.51}$$

where

$$\Delta T = (T_w - T_e) \tag{4.52}$$

and

$$\eta = \frac{y}{\delta} \tag{4.53}$$

The characteristic velocity U_c and the temperature difference ΔT are assumed to be functions of x. Further, defining the ratio of the buoyancy force to the shear force parameter $\widehat{\xi}$, as

$$\widehat{\xi} = \rho g_x \beta_0 \int_0^{\delta} (T - T_e) dy / \tau_w \tag{4.54}$$

The governing Eqs. (4.42) and (4.43) using Eq. (4.49) are written as

$$C_0 \frac{d}{dx} r^* U_c^2 \delta = \left(\widehat{\xi} - 1\right) C_1 r^* g_x \beta_0 \Delta T \delta / \widehat{\xi} \tag{4.55}$$

$$C_2 \frac{d}{dx} r^* U_c C_p \Delta T \delta = r^* \tau_w \Delta T \mathrm{Pr}_c^{-\frac{2}{3}} / \rho U_c \tag{4.56}$$

where

$$C_0 = \int_0^1 u_1^2 d\eta \tag{4.57a}$$

$$C_1 = \int_0^1 \theta d\eta \tag{4.57b}$$

$$C_2 = \int_0^1 u_1 \theta d\eta \tag{4.57c}$$

Based on the right-hand side of Eq. (4.55), it is evident that the parameter $\widehat{\xi}$ directly governs the flow acceleration and that the condition $\widehat{\xi}(x) = 1$ must be satisfied everywhere for the flow to be thermally stable. Before solving the above equations, it is necessary to get an expression for τ_w for the free convection turbulent flow under consideration. Drag reducing fluids are known to be Newtonian in viscosity but exhibit mild elasticity characterized by a relaxation time θ_{fl}. Detailed discussions on the determination of relaxation times for drag reducing fluids are available (Argumedo, Tung, & Chang, 1978; Cho & Hartnett, 1982). For such fluids, one assumes the friction factor f to be a function of the Reynolds number Re and the Deborah number De (which is the ratio of the fluid relaxation time θ_{fl} and the characteristic process time v/u^{*2}).

The Blasius-type correlation developed (Shenoy & Mashelkar, 1983) for drag reducing fluids gives the following form for friction factor:

$$f = \frac{\alpha}{\mathrm{Re}^{\beta}} \text{ for } 5 \times 10^3 \leq \mathrm{Re} \leq 10^5 \text{ and } 0 \leq \mathrm{De} \leq 10 \tag{4.58}$$

α and β are both functions of Deborah number, and their values for varying De are given (Shenoy & Shintre, 1986) and shown in Table 2.1.

It has been suggested (Virk, 1966) that the maximum drag reduction that can be achieved in practice can be described by a unique asymptote given by $\alpha = 0.42$ and $\beta = 0.55$ in Eq. (2.25). In Table 2.1, these values correspond to the limit $\mathrm{De} \geq 20$ which has been used to denote the maximum drag reduction asymptote.

As can be seen, the value of α is almost constant for $1 \leq \mathrm{De} \leq 10$ but jumps by about 600% for a change of De from 10 to 20. This is due to the fact that around $\mathrm{De} = 20$ and beyond it, the f versus Re curve has a sudden change of slope, and hence it is not only the value of α but also the value of β that undergoes a sudden change. Detailed explanation of the maximum drag reduction asymptote and its uniqueness is available elsewhere (Virk, 1975) in the exhaustive article and can be referred to for details. Equation (4.58) has been used when analyzing turbulent flow in horizontal pipes (Shenoy & Mashelkar, 1983), in curved tubes (Shenoy, Ranade, & Ulbrecht, 1980), in rotating straight tubes (Shenoy, 1986), in annular ducts (Shenoy & Shintre, 1986), and in vertical tubes (Shenoy, 1987).

A suitable expression for the local surface shear stress can be obtained by proper rearranging and adapting Eq. (4.58) to flow over a smooth flat plate at zero incidence in a manner analogous to that used in the Newtonian case (Eckert & Jackson, 1950) as

$$\frac{\tau_w}{\rho U_c^2} = \Omega \left(\frac{\mu}{\rho U_c \delta} \right)^{\beta} \tag{4.59}$$

where

$$\Omega = \frac{\alpha(0.817)^{2-\beta}}{2^{\beta+1}} \tag{4.60}$$

Note that for the Newtonian case

$$\beta = 0.25; \quad \Omega = 0.02332 \tag{4.61}$$

$$\frac{\tau_w}{\rho U_c^2} = 0.02332\left(\frac{\mu}{\rho U_c \delta}\right)^{\frac{1}{4}} \tag{4.62}$$

Substitution of Eq. (4.59) into Eqs. (4.55) and (4.56) gives the final simplified forms of the governing equations that are now to be solved. This requires expressions for the dimensionless velocity and temperature profiles. These are sought by following the arguments set forth for Newtonian fluids (Eckert & Jackson, 1950). They noted that in turbulent forced convection equations of the form $u_1(\eta) = \eta^{1/7}$ and $\theta(n) = 1 - \eta^{1/7}$ hold rather well. For turbulent free convection, they found that experimental data could be fitted well with the same equation for temperature profile, while the velocity profile needed to be modified to $u_1(\eta) = \eta^{1/7}(1 - \eta)^4$.

For drag reducing fluids, the velocity profile for turbulent forced convection flow can be taken as $u_1(\eta) = \eta^q$ where $q = \beta/(2 - \beta)$. In the free convection case, the velocity and temperature profiles will be assumed by analogous arguments to those for Newtonian fluids (Eckert & Jackson, 1950), making use of the forced convection expression for drag reducing fluids as stated above. Thus, the dimensionless velocity and temperature profiles that are assumed to fit the turbulent free convection flow of drag reducing fluids are

$$u_1(\eta) = \eta^q(1-\eta)^4 \tag{4.63}$$

$$\theta_1(\eta) = 1 - \eta^q \tag{4.64}$$

where

$$q = \frac{\beta}{2-\beta} \tag{4.65}$$

Using Eqs. (4.63) and (4.64), the expressions for C_0, C_1, and C_2 as defined in Eqs. (4.57a), (4.57b), and (4.57c) can be easily obtained as

$$C_0 = \frac{1}{2q+1} - \frac{4}{q+1} + \frac{28}{2q+3} - \frac{28}{q+2} + \frac{70}{2q+5} - \frac{28}{q+3} + \frac{28}{2q+7} - \frac{4}{q+4} + \frac{1}{2q+9} \tag{4.66}$$

$$C_1 = \frac{q}{q+1} \tag{4.67}$$

$$C_2 = \frac{3}{q+1} - \frac{2}{q+2} + \frac{6}{q+3} - \frac{4}{q+4} + \frac{1}{q+5} - \frac{1}{2q+1} - \frac{6}{2q+3} - \frac{1}{2q+5} \tag{4.68}$$

Note that for Newtonian fluids when q takes the value of 1/7, $C_0 = 0.0523$, $C_1 = 1/8$, and $C_2 = 0.0366$, which are all identical to the values obtained for Newtonian fluids (Nakayama & Koyama, 1985). Upon combining Eqs. (4.54) and (4.59), one obtains the following expression for U_c:

$$U_c = \left(\frac{C_1 g_x \beta_0 \Delta T}{\Omega \nu^\beta \widehat{\xi}} \right)^{\frac{1}{2-\beta}} \delta^{\frac{1+\beta}{2-\beta}} \tag{4.69}$$

where ν is the kinematic viscosity. This equation along with Eq. (4.59) can now be substituted into the governing Eqs. (4.55) and (4.56) to eliminate U_c and τ_w and thus results in the following equations after mathematical rearrangement of the terms:

$$\frac{d}{dx}\delta^{\frac{2(1+\beta)}{2-\beta}} + \delta^{\frac{2(1+\beta)}{2-\beta}} \frac{d}{dx} \ln \left[r^* \Delta T \left(\frac{g_x \Delta T}{\widehat{\xi}} \right)^{\frac{2}{2-\beta}} \right]^{\frac{2(1+\beta)}{4+\beta}} = \frac{2(1+\beta)}{4+\beta} \frac{C_1}{C_0} \left(\frac{\Omega}{C_1} \right)^{\frac{2}{2-\beta}} \left(\widehat{\xi} - 1 \right) \left(\frac{\widehat{\xi} \nu^2}{g_x \beta_0 \Delta T} \right)^{\frac{\beta}{2-\beta}} \tag{4.70}$$

$$\frac{d}{dx}\delta^{\frac{2(1+\beta)}{2-\beta}} + \delta^{\frac{2(1+\beta)}{2-\beta}} \frac{d}{dx} \ln \left[r^* \Delta T \left(\frac{g_x \Delta T}{\widehat{\xi}} \right)^{\frac{1}{2-\beta}} \right]^{\frac{2(1+\beta)}{3}} = \frac{2(1+\beta)}{3} \frac{\Omega}{C_2} \left(\frac{\Omega}{C_1} \right)^{\frac{\beta}{2-\beta}} \mathrm{Pr}_c^{-\frac{2}{3}} \left(\frac{\widehat{\xi} \nu^2}{g_x \beta_0 \Delta T} \right)^{\frac{\beta}{2-\beta}} \tag{4.71}$$

Integration of the above set of equations yields two distinct expressions as follows:

$$\left(\frac{\delta}{x} \right)^{\frac{2(1+\beta)}{2-\beta}} \mathrm{Gr}_x^{\frac{\beta}{2-\beta}} = \frac{2(1+\beta)}{4+\beta} \frac{C_1}{C_0} \left(\frac{\Omega}{C_1} \right)^{\frac{2}{2-\beta}} \left(\widehat{\xi} - 1 \right) \left(\widehat{\xi} \right)^{\frac{\beta}{2-\beta}} I_v \tag{4.72}$$

and

$$\left(\frac{\delta}{x}\right)^{\frac{2(1+\beta)}{2-\beta}} \mathrm{Gr}_{\mathrm{x}}^{\frac{\beta}{2-\beta}} = \frac{2(1+\beta)}{3} \frac{\Omega}{C_2} \left(\frac{\Omega}{C_1}\right)^{\frac{\beta}{2-\beta}} \mathrm{Pr}_{\mathrm{c}}^{-\frac{2}{3}} \left(\widehat{\xi}\right)^{\frac{\beta}{2-\beta}} I_{\mathrm{t}} \tag{4.73}$$

where

$$\mathrm{Gr}_{\mathrm{x}} = \frac{g_{\mathrm{x}} \beta_0 \Delta T x^3}{\nu^2} \tag{4.74}$$

is the local Grashof number and

$$I_{\mathrm{v}} = \frac{\int_0^x \left(\widehat{\xi} - 1\right) \left[r^{*2(1+\beta)} \left(\frac{g_{\mathrm{x}} \Delta T}{\widehat{\xi}}\right)^{2+\beta} \right]^{\frac{1}{4+\beta}} dx}{x\left(\widehat{\xi} - 1\right) \left[r^{*2(1+\beta)} \left(\frac{g_{\mathrm{x}} \Delta T}{\widehat{\xi}}\right)^{2+\beta} \right]^{\frac{1}{4+\beta}}} \tag{4.75}$$

$$I_{\mathrm{t}} = \frac{\int_0^x \left(r^* \Delta T\right)^{\frac{2(1+\beta)}{3}} \left(\frac{g_{\mathrm{x}} \Delta T}{\widehat{\xi}}\right)^{\frac{1}{3}} dx}{x\left(r^* \Delta T\right)^{\frac{2(1+\beta)}{3}} \left(\frac{g_{\mathrm{x}} \Delta T}{\widehat{\xi}}\right)^{\frac{1}{3}}} \tag{4.76}$$

The foregoing functions I_{v} and I_{t} account for the total combined effects of arbitrary geometries and wall temperature distributions. The local Nusselt number Nu_{x}, which is of primary interest, is related to U_{c} and δ via the Colburn analogy as

$$\mathrm{Nu}_{\mathrm{x}} = \frac{hx}{k} = \mathrm{Pr}_{\mathrm{c}}^{\frac{1}{3}} \left(\frac{U_{\mathrm{c}} x}{\nu}\right) \left(\frac{\tau_{\mathrm{w}}}{\rho U_{\mathrm{c}}^2}\right) = \mathrm{Pr}_{\mathrm{c}}^{\frac{1}{3}} \Omega \left(\frac{U_{\mathrm{c}} x}{\nu}\right)^{1-\beta} \left(\frac{x}{\delta}\right)^{\beta} \tag{4.77}$$

where h and k are the local heat transfer coefficient and thermal conductivity. U_{c} in the above equation may be eliminated in favor of δ, using Eq. (4.69). After substitution of Eq. (4.73) into the equation and after some manipulation, the Nu_{x} expression is obtained as follows:

$$\mathrm{Nu}_{\mathrm{x}} = \left\{\frac{2(1+\beta) I_{\mathrm{t}}}{3 C_2}\right\}^{\frac{1-2\beta}{2(1+\beta)}} \left(\Omega C_1^{\frac{1}{2}}\right)^{\frac{1}{1+\beta}} \mathrm{Pr}^{\frac{\beta}{1+\beta}} \left(\frac{\mathrm{Gr}_{\mathrm{x}}}{\widehat{\xi}}\right)^{\frac{1}{2(1+\beta)}} \tag{4.78}$$

Having established all the necessary relations, the solution of the problem is now reduced to the determination of the unknown acceleration parameter $\widehat{\xi}(x)$. For this purpose, the two distinct expressions for δ, namely, Eqs. (4.72) and (4.73), are equated to give the following characteristic equation:

$$\widehat{\xi} = 1 + \frac{C_0(4+\beta)I_t}{3C_2I_v}\mathrm{Pr_c}^{-\frac{2}{3}} \tag{4.79}$$

Since Eqs. (4.75) and (4.76) for I_v and I_t also involve the unknown $\widehat{\xi}(x)$, the foregoing characteristic equation is implicit in $\widehat{\xi}$. Thus, the determination of $\widehat{\xi}(x)$ in general requires an iterative procedure at each integration step. A simple way to find $\widehat{\xi}(x)$ is to guess $\widehat{\xi}$ at the end of each integration step, and evaluate Eqs. (4.75) and (4.76), using this guessed value and the value determined during the preceding integration step (corresponding to the value at the beginning of the current integration step). The integration results are to be substituted back into the characteristic Eq. (4.79) to check if the estimated $\widehat{\xi}$ at the end of the integration step satisfies the relation. This sequence must be repeated to determine $\widehat{\xi}$ within a desired accuracy before marching one step further. The boundary value $\widehat{\xi}(0)$ needed for initiation of such integrations, however, must be provided prior to the downstream marching in consideration of the similarity solutions presented in the following section. Certain cases for which the functions I_v and I_t remain constant indicate the existence of similarity solutions. Any geometry near the stagnation point may be specified as

$$r = \left(\left.\frac{dr}{dx}\right|_{x=0}\right)x \tag{4.80}$$

Hence,

$$\cos\emptyset = \left\{1 - \left(\left.\frac{dr}{dx}\right|_{x=0}\right)^2\right\}^{\frac{1}{2}} = \text{constant for pointed body} \tag{4.81}$$

$$r = r_0 \sin\left(\frac{x}{r_0}\right) \tag{4.82}$$

$$\cos\emptyset = \sin\left(\frac{x}{r_0}\right) = \text{constant for blunt body} \tag{4.83}$$

where r_0 is the local radius at $x = 0$. The foregoing consideration reveals the following proportional relationship:

$$r \propto x^i \text{ where } i = 0 \text{ for plane body and } i = 1 \text{ for axisymmetric body} \tag{4.84}$$

$$g_x \propto x^j \text{ where } j = 0 \text{ for pointed body and } i = 1 \text{ for blunt body} \tag{4.85}$$

For example, the integers (i, j) should be set to (0, 0) for a flat plate, (1, 0) for a vertical cone pointing downward, (0, 1) for the stagnation region on a horizontal circular cylinder, and (1, 1) for the stagnation region of a sphere. Equations (4.75) and (4.76) under the conditions described by the foregoing proportional relationships yield

$$I_{\mathrm{v}} = \left\{1 + \frac{2(1+\beta)i + (2+\beta)(j+m_{\mathrm{t}})}{4+\beta}\right\}^{-1} \tag{4.86}$$

$$I_{\mathrm{t}} = \left\{1 + \frac{2(1+\beta)i}{3} + \frac{j}{3} + \frac{(3+2\beta)m_{\mathrm{t}}}{3}\right\}^{-1} \tag{4.87}$$

where m_{t} is associated with the wall temperature distribution around the stagnation point, which is assumed to follow

$$\Delta T \propto x^{m_{\mathrm{t}}} \tag{4.88}$$

The wall temperature distribution effect on the functions I_{v} and I_{t} is such that both I_{v} and I_{t} diminish as ΔT_{w} increases downstream (i.e., $m_{\mathrm{t}} > 0$). Substitution of Eqs. (4.86) and (4.87) into the characteristic Eq. (4.79) gives an explicit equation for $\widehat{\xi}$:

$$\widehat{\xi} = 1 + \frac{C_0(4+\beta)}{3C_2}\frac{\left\{1 + \frac{2(1+\beta)i + (2+\beta)(j+m_{\mathrm{t}})}{4+\beta}\right\}}{\left\{1 + \frac{2(1+\beta)i}{3} + \frac{j}{3} + \frac{(3+2\beta)m_{\mathrm{t}}}{3}\right\}\mathrm{Pr}_{\mathrm{c}}^{\frac{2}{3}}} \tag{4.89}$$

Thus, $\widehat{\xi}$ stays constant around $x = 0$, and a similarity solution exists around the stagnation point. For the isothermal vertical flat plate in contact with Newtonian fluids (i.e., $\beta = 0.25$ and $i = j = m_{\mathrm{t}} = 0$), both I_{v} and I_{t} become unity, and

$$\widehat{\xi} = 1 + 2.023\mathrm{Pr}_{\mathrm{c}}^{-\frac{2}{3}} \tag{4.90}$$

When the above equation is substituted into Eq. (4.78), it reduces to the form derived in the case of Newtonian fluids (Eckert & Jackson, 1950) for an isothermal vertical flat plate, namely,

$$\mathrm{Nu}_{\mathrm{x}} = \frac{0.0402\mathrm{Pr}^{\frac{1}{3}}}{\left(1 + 2.023\mathrm{Pr}_{\mathrm{c}}^{-\frac{2}{3}}\right)^{\frac{2}{5}}}\mathrm{Gr}_{\mathrm{x}}^{\frac{2}{5}} \tag{4.91}$$

For the drag reducing fluids, the expressions for the acceleration parameter $\widehat{\xi}$ and the local Nusselt number Nu_{x} are given as

$$\widehat{\xi} = 1 + \overline{C_2}\frac{I_{\mathrm{t}}}{I_{\mathrm{v}}}\mathrm{Pr}_{\mathrm{c}}^{-\frac{2}{3}} \tag{4.92}$$

and

$$\mathrm{Nu_x} = \frac{\overline{C_1} I_t^{a''-b''} \mathrm{Pr}^{b''}}{\left(1 + \overline{C_2}\left(\frac{I_t}{I_v}\right) \mathrm{Pr_c}^{-\frac{2}{3}}\right)^{a''}} = \overline{C_3} \mathrm{Gr_x}^{a''} \tag{4.93}$$

where

$$a'' = \frac{1}{2(1+\beta)} \tag{4.94}$$

$$b'' = \frac{\beta}{1+\beta} \tag{4.95}$$

$$\overline{C_1} = \left\{\frac{2(1+\beta)}{3C_2}\right\}^{\frac{1-2\beta}{2(1+\beta)}} \left(\Omega C_1^{\frac{1}{2}}\right)^{\frac{1}{1+\beta}} \tag{4.96}$$

$$\overline{C_2} = \frac{C_0(4+\beta)}{3C_2} \tag{4.97}$$

For the similarity solutions, I_v and I_t, in the above expression (4.93), should be provided according to Eqs. (4.86) and (4.87). The numerical values of a'', b'', $\overline{C_1}$, and $\overline{C_2}$ are furnished in Table 4.2, as function of De.

The effects of De and Pr on the acceleration parameter are illustrated in Fig. 4.3, for the case of the isothermal vertical flat plate. As expected from Eq. (4.92), increase in either De or Pr results in decrease in ξ. It should be noted that ξ decreases down to unity as Pr goes to infinity, and the inertia effects vanish totally. The values for the multiplicative constant $\overline{C_3}$ (as a function of both De and Pr) were calculated assuming Pr = 7 and listed in Table 4.3 for the isothermal flat plate and cone as well as the stagnation regions of isothermal cylinder and sphere.

Table 4.2 Values of a'', b'', $\overline{C_1}$, and $\overline{C_2}$ for varying values of Deborah numbers De for turbulent free convection from a vertical flat plate (Nakayama & Shenoy, 1992)

De	α	β	a''	b''	$\overline{C_1}$	$\overline{C_2}$
0	0.079	0.250	0.400	0.200	0.0402	2.02
1	0.0782	0.262	0.396	0.208	0.0404	1.91
2	0.0787	0.271	0.393	0.213	0.0401	1.82
3	0.0741	0.278	0.391	0.218	0.0393	1.76
4	0.0726	0.285	0.389	0.222	0.0390	1.71
5	0.0689	0.289	0.388	0.224	0.0375	1.67
6	0.0655	0.292	0.387	0.226	0.0382	1.65
7	0.0662	0.301	0.384	0.231	0.0368	1.59
8	0.0687	0.312	0.381	0.238	0.0383	1.51
9	0.0732	0.324	0.378	0.245	0.0406	1.43
10	0.0762	0.334	0.375	0.250	0.0422	1.38
≥20	0.420	0.550	0.323	0.355	0.141	0.623

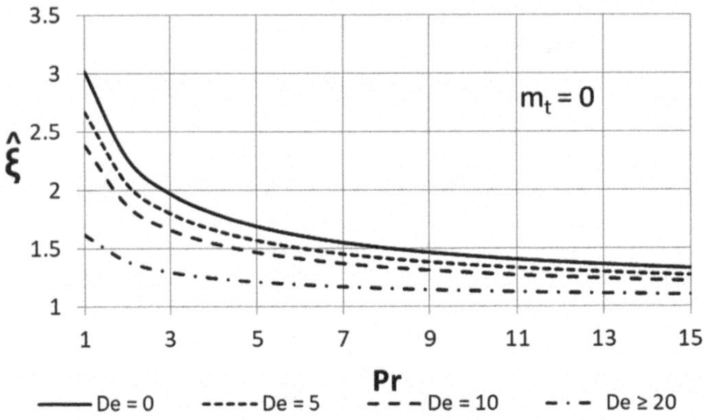

Fig. 4.3 Effects of Pr and De on $\hat{\xi}$ for isothermal vertical flat plate (Nakayama & Shenoy, 1992). (Reprinted with permission from ASME)

Table 4.3 Values of coefficient $\overline{C_3}$ for $Pr = 7$ with varying values of Deborah number for turbulent free convection from isothermal bodies (Nakayama & Shenoy, 1992)

De	Flat plate (0, 0)	Cone (1, 0)	Cylinder (0, 1)	Sphere (1, 1)
0	0.0498	0.0450	0.0460	0.0428
1	0.0512	0.0465	0.0478	0.0444
2	0.0518	0.0472	0.0483	0.0452
3	0.0515	0.0471	0.0481	0.0451
4	0.0517	0.0474	0.0484	0.0455
5	0.0502	0.0481	0.0470	0.0443
6	0.0487	0.0448	0.0456	0.0430
7	0.0503	0.0465	0.0473	0.0447
8	0.0533	0.0495	0.0503	0.0478
9	0.0577	0.0539	0.0547	0.0521
10	0.0610	0.0572	0.0579	0.0554
≥20	0.287	0.275	0.287	0.275

Reprinted with permission from ASME

Table 4.3 shows that $\overline{C_3}$ is nearly constant for $De \leq 7$, and thereafter increases, jumping to a high level at $De = 20$. Within the range of $De \leq 7$, the $\overline{C_3}$ value goes up slightly and then comes down slightly. This "up-and-down" variation in $\overline{C_3}$ should not be considered to be serious, since it is not the slight change in the $\overline{C_3}$ value, but the exponent $a''(=1/2(1+\beta))$ that virtually determines the level of Nu_x when Gr_x is large as in the present case of turbulent natural convection. It is to be noted that each set of a″ and $\overline{C_3}$ for $De < 10$ follows the expected trend, namely, that increase in De results in decrease in Nu_x, as shown in Fig. 4.4, for the case of isothermal vertical flat plate.

The figure clearly shows that the addition of small amounts of drag reducing polymer results in reduction in heat transfer as well. This fact, which is well-known

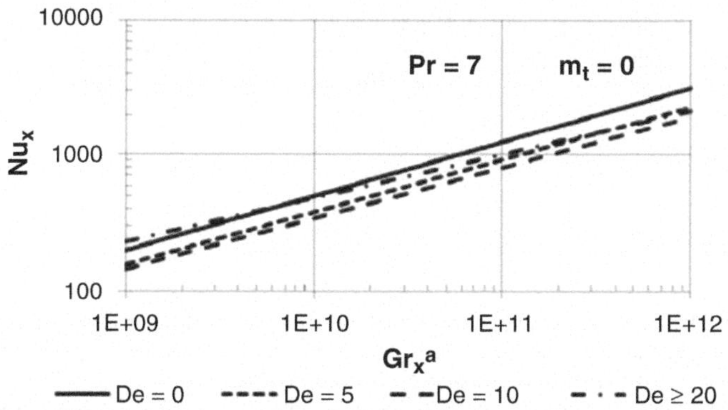

Fig. 4.4 Variation of Nusselt number Nu_x with Grashof number Gr_x at $Pr = 7$ and various selected values of De for isothermal vertical flat plate (Nakayama & Shenoy, 1992). (Reprinted with permission from ASME)

for turbulent forced convection heat transfer to drag reducing fluids, holds true for turbulent free convection as well. In Fig. 4.4, the asymptote corresponding to maximum drag reduction is indicated by the dashed-dotted line. It is interesting to note that, within the Gr_x range considered herein, the heat transfer rate for $De = 10$ is found to be even lower than the rate for the case of the maximum drag reduction. This is not an anomaly because it is known that there is a delay in the onset of drag reduction with increasing drag reducing efficiency. This delay is not pronounced at lower De and hence is not noticeable for $De \leq 10$ within the considered Gr_x range. However, for the maximum drag reduction asymptote, the onset of drag reduction is delayed to a predominant extent and hence the heat transfer rate departure from the Newtonian case would also be delayed to a great extent. It is worth noting that the maximum drag reduction asymptote will, undoubtedly, result in the maximum reduction in the heat transfer rate, but this will occur at much higher Gr_x beyond what is covered in Fig. 4.4.

There is evidence in the literature (Cho & Hartnett, 1982; Marrucci & Astarita, 1967; Mizushina & Usui, 1977; Ng, Cho, & Hartnett, 1980; Tsukahara & Kawaguchi, 2011) that the heat transfer reduction is comparatively larger than the drag reduction during turbulent flow through smooth circular pipes. This is likely to hold good even for external flow situations though analogy with pipe flow may not be true as mechanisms for internal and external flow are not quite the same. However, it must be borne in mind that the analysis in this section was initiated through a scaling of the terms of the governing equations resulting in Eq. (4.49), which implied that the heat and momentum transfer are of the same order. Hence, the results of the analysis may underpredict the heat transfer reductions to a certain extent.

The averaged Nusselt number Nu_L, which is often more convenient to use for heat transfer estimation, can easily be derived by taking an integrated average over length L:

$$Nu_{L_{av}} = \frac{h_{av}L}{k} = \frac{L\int_0^L \left(\frac{Nu_x}{x}\right) r^* dx}{\int_0^L r^* dx} = \frac{1+i}{i + \frac{3+j+m_t}{2(1+\beta)}} Nu_x \Bigg|_{x=L} \tag{4.98}$$

References

Argumedo, A., Tung, T. T., & Chang, K. I. (1978). Rheological property measurements of drag reducing polyacrylamide solutions. *Transactions Society of Rheology, 22*, 449.

Bejan, A. (1984). *Convective heat transfer*. New York: Wiley.

Cho, Y. I., & Hartnett, J. P. (1982). Non-newtonian fluids in circular pipe flows. *Advances Heat Transfer, 15*, 59–141.

Eckert, E. R., & Jackson, T. (1950). *Analysis of turbulent free convection boundary layer on a flat plate* (p. 2207). Washington, DC: National Advisory Committee of Aeronautics Technology Note.

Griffiths, E., & Davis, A. H. (1922). *The transmission of heat by radiation and convection* (Report No. 9). DSIR-Food Invest. Bd. Spec.

Hellums, J. D., & Churchill, S. W. (1964). Simplification of the mathematical description of boundary and initial value problems. *AICHE Journal, 10*(1), 110–114.

Marrucci, G., & Astarita, G. (1967). Turbulent heat transfer in viscoelastic liquids. *Industrial and Engineering Chemistry Fundamentals, 6*(3), 470–471.

Mizushina, T., & Usui, H. (1977). Reduction of eddy diffusion for momentum and heat in viscoelastic fluid flow in a circular tube. *Physics of Fluids, 20*(10), S100–S108.

Nakayama, A., & Koyama, H. (1985). An analysis of turbulent free convection about bodies of arbitrary geometrical configurations. *Warme-und Stoffuberttragung, 19*, 263–268.

Nakayama, A., & Shenoy, A. V. (1992). Turbulent free convection heat transfer to drag-reducing fluids from arbitrary geometric configurations. *Transactions ASME Journal of Heat Transfer, 114*(1), 127–134.

Ng, K. S., Cho, Y. I., & Hartnett, J. P. (1980). Heat transfer performance of concentrated polyethylene oxide and polyacrylamide solutions. *AICHE Symposium Series No. 199, 76*, 250–256.

Shenoy, A. V. (1986). Turbulent flow of mildly elastic fluids through rotating straight circular tubes. *Journal of Applied Sciences Research, 43*(1), 39–54.

Shenoy, A. V. (1987). Effects of bouyancy on heat transfer during turbulent flow of drag reducing fluids in vertical pipes. *Warme- und Stoffubertragung, 21*(1), 15–18.

Shenoy, A. V., & Mashelkar, R. A. (1983). Engineering estimate of hydrodynamic entrance lengths in non-newtonian turbulent flow. *Industrial and Engineering Chemistry Process Design and Development, 22*(1), 165–168.

Shenoy, A. V., & Shintre, S. N. (1986). Developing and fully developed turbulent flow of drag reducing fluids in an annular duct. *The Canadian Journal of Chemical Engineering, 64*(2), 190–195.

Shenoy, A. V., Ranade, V. R., & Ulbrecht, J. J. (1980). Turbulent flow of mildly viscoelastic liquids in curved tubes. *Chemical Engineering Communications, 5*(5–6), 269–286.

Skelland, A. H. (1967). *Non-newtonian flow and heat transfer*. New York: Wiley.

Tsukahara, T., & Kawaguchi, Y. (2011). Turbulent heat transfer in Drag-Reducing Channel flow of viscoelastic fluid. In A. Ahsan (Ed.), *Evaporation, condensation and heat transfer* (pp. 375–400). Croatia, Balkans. *(www.intechopen.com)*: InTech.

Virk, P. S. (1966). *The toms phenomenon – Turbulent pipe flow of dilute polymer solutions*, Mass. Inst. of Tech., (ScD Thesis).

Virk, P. S. (1975). Drag reduction fundamentals. *AICHE Journal, 21*(4), 625–656.

Chapter 5
Turbulent Forced and Mixed Convection Heat Transfer in Internal Flows

Momentum/Heat Transfer Analogy

It is well-known that the heat transfer rate can be reasonably well estimated without actually solving the energy equation through the use of momentum/heat transfer analogies. For Newtonian fluids, the alternative approaches for establishing the momentum and heat transfer during turbulent flow have been discussed, and the momentum/heat transfer analogy for external turbulent boundary-layer flow under the influence of mild pressure gradients has been derived (Nakayama, Koyama, & Ohsawa, 1984).

Due to the tremendous interest in the field of drag reduction, there have been various attempts to develop momentum, heat, and mass transfer analogies for drag reducing fluids as evidenced by the efforts that have gone into this area of research (Cho & Hartnett, 1980; Dudukovic, 1988; Friend & Metzner, 1958; Gupta, Metzner, & Hartnett, 1967; Kale, 1977; Kawase & Ulbrecht, 1982; Metzner & Friend, 1959; Poreh & Paz, 1968; Smith, Keuroghlian, Virk, & Merrill, 1969; Wells, 1968).

Stanton number as a function of the friction factor and Prandtl number was calculated (Metzner & Friend, 1959), applying Riechardt's general formulation for the analogy between heat and momentum transfer in turbulent flow. The correlation gave fairly good predictions for purely viscous fluids, but not for mildly elastic drag reducing fluids. The model proposed later (Poreh & Paz, 1968) was based on the assumption that the eddy viscosity can be taken to be identically equal to zero in the laminar sublayer and, hence, their predictions are inaccurate at high Prandtl numbers. It was shown (Cho & Hartnett, 1980) that there is no simple and direct analogy between momentum, heat, and mass transfer in mildly elastic drag reducing fluids; however, this was not acceptable (Dudukovic, 1988). In fact, a general type of analogy was proposed (Dudukovic, 1988) which contained both the laminar and the turbulent Prandtl number. The laminar Prandtl number would take into account different relative contributions of molecular transport, while the turbulent Prandtl number represents the changes in the turbulence spectra and their effect on the rate or

A. Shenoy, *Rheology of Drag Reducing Fluids*,
https://doi.org/10.1007/978-3-030-40045-3_5

the momentum, heat, and mass transfer processes. Effects of buoyancy on heat transfer and temperature profiles in horizontal pipe flow of drag reducing fluids have been studied (Gasljevic, Aguilar, & Matthys, 2000).

In the present section, the solution method for Newtonian fluids (Nakayama et al., 1984) is used to analyze the turbulent boundary layer in the presence of small pressure gradients in order to establish the momentum/heat transfer analogy for mildly elastic drag reducing fluids (Shenoy, 1992).

The expression for the velocity profile in the presence of a small pressure gradient for drag reducing fluids can be derived in a manner analogous to that for Newtonian fluids (Nakayama et al., 1984) and written as follows:

$$\frac{u}{\left(\frac{\tau_w}{\rho}\right)^{0.5}} = \left(\frac{1}{\kappa}\right)\ln\left(\frac{y}{y_x}\right) + \left(\frac{1}{\kappa'}\right)\sum_{m'=1}^{\infty}\binom{\frac{1}{2}}{m'}\left(\frac{\beta_1^{m'}}{m'}\right)\left(\frac{y^{m'} - y_s^{m'}}{\delta^{m'}}\right) \tag{5.1}$$

where

$$\binom{\frac{1}{2}}{m'} = \frac{1}{2}\left(\frac{1}{2} - 1\right)\left(\frac{1}{2} - 2\right)\ldots\left(\frac{1}{2} - m' + 1\right)/m'! \tag{5.2}$$

and where u is the mean velocity in the streamwise direction x and τ is the local shear stress at the normal distance y away from the wall. The subscript w refers to the wall. For Newtonian fluids, κ' is the von Karman constant, and for drag reducing fluids, this will be derived later from the known velocity profile for drag reducing fluids.

The stress gradient is assumed to be small enough such that $|\beta_1 y/\delta| < 1$, where

$$\beta_1 = \frac{\delta}{\tau_w}\left(\frac{d\tau}{dy}\right)_w \tag{5.3}$$

For drag reducing fluids flowing through a pipe, an expression for the velocity profile has been provided (Shenoy & Talathi, 1985) which can be easily adapted to boundary-layer flow situation to give the following:

$$u^+ = A' \ln y^+ + (A'C'' + B'') \tag{5.4}$$

where

$$A' = 2.46 \tag{5.5}$$

$$B'' = 1.6 + 1.55\text{De} \tag{5.6}$$

$$C'' = \left(0.4398 + 0.123\mathrm{De} + 0.0135\mathrm{De}^2\right)\exp\left[-4.961(1 + 0.2\mathrm{De})^2\right] + 1.3676\left(1 - 0.09\mathrm{De} - 0.01\mathrm{De}^2\right) \tag{5.7}$$

$$u^+ = \frac{u}{\left(\frac{\tau_w}{\rho}\right)^{0.5}} \tag{5.8}$$

$$y^+ = \frac{y\left(\frac{\tau_w}{\rho}\right)^{0.5}}{\nu} \tag{5.9}$$

Comparison of Eq. (5.1) with Eq. (5.4) for $\beta_1 = 0$ implies that

$$\kappa' = \frac{1}{A'} = 0.4065 \tag{5.10}$$

$$\frac{y_s\left(\frac{\tau_w}{\rho}\right)^{0.5}}{\nu} = e^{\frac{-(A'C'' + B'')}{A'}} \tag{5.11}$$

Equation (5.11) is evaluated for different values of De as given in Table 5.1.

Following the arguments given for Newtonian fluids (Nakayama et al., 1984), it is assumed that the temperature law of the wall for zero pressure gradient given below may well be valid even for the case of small pressure gradients:

$$\frac{\rho C_p\left(\frac{\tau_w}{\rho}\right)^{0.5}(T_w - T)}{q_w} = A\ln\left(\frac{y}{y_s}\right) + P \tag{5.12}$$

Table 5.1 Values of expression exp $[-(A'C'' + B'')/A']$ for different values of De, α, and β (Shenoy, 1992)

De	α	β	exp $[-(A'C'' + B'')/A']$
0	0.079	0.250	0.1325
1	0.0782	0.262	0.0808
2	0.0787	0.271	0.0506
3	0.0741	0.278	0.0325
4	0.0726	0.285	0.0214
5	0.0689	0.289	0.0145
6	0.0655	0.292	0.1006
7	0.0662	0.301	0.0072
8	0.0687	0.312	0.0052
9	0.0732	0.324	0.0039
10	0.0762	0.334	0.0030
$\geq$20	0.420	0.550	0.0006

where P is the "P-function" (Jayatillaka, 1969) that accounts for the enhanced resistance to heat transfer offered by the viscous sublayer as a function of laminar Prandtl number Pr. Using a large amount of experimental values from the available literature on Newtonian fluids, the following simple form for the P-function was given (Jayatillaka, 1969) which predicted the extra resistance to heat transfer rather accurately:

$$P = 9.24\left(\mathrm{Pr}^{\frac{3}{4}} - 1\right) \tag{5.13}$$

For mildly elastic drag reducing fluids, the same procedure (Jayatillaka, 1969) could be followed for derivation of the P-function. However, a lot of accurate flow and heat transfer data on drag reducing fluids is required. Though there is a lot of available heat transfer data in the literature on drag reducing fluids, there is a rather wide scatter in the data (Gasljevic et al., 2000; Ng, Cho, & Hartnett, 1980) due to various reasons such as thermal entrance effects, degradation of test fluids, large bulk to wall temperature difference, and solvent effect. When reliability of the data is under question, it is simply not worth trying to find an equation similar to Eq. (5.13) by the use of experimental findings. Hence, under such circumstances, it is assumed that the form of Eq. (5.13) for Newtonian fluids (Jayatillaka, 1969) holds for drag reducing fluids as well, as a first approximation. Note that in Eqs. (5.12) and (5.13), the turbulent Prandtl number is assumed to be unity. After evaluating Eqs. (5.1) and (5.12) at the viscous ($y = \delta$) and the thermal ($y = \delta_T$) boundary-layer edges, respectively, the subtraction of Eq. (5.12) from Eq. (5.1) leaves the following:

$$\left(\frac{2}{C_{fx}}\right)^{\frac{1}{2}} - \frac{\left(\frac{C_{fx}}{2}\right)^{\frac{1}{2}}}{S_{tx}} = A \ln\left(\frac{\delta}{\delta_T}\right) - P + A\sum\nolimits_{m'=1}^{\infty}\begin{pmatrix}\frac{1}{2}\\ m'\end{pmatrix}\left(\frac{\beta_1^{m'}}{m'}\right) \times \left[1 - \left(\frac{y_s^{m'}}{\delta^{m'}}\right)\right] \tag{5.14}$$

where the skin friction coefficient is

$$C_{fx} = \frac{2\tau_w}{\rho u_e^2} \tag{5.15}$$

and the Stanton number

$$S_{tx} = \frac{q_w}{\rho C_p u_e (T_{w-}T_e)} \tag{5.16}$$

The subscript e refers to the corresponding boundary-layer edge $y = \delta$ or δ_T. Due to Eq. (5.11), (y_s/δ) in the last term of the right-hand side of Eq. (5.14) may be dropped. Moreover, the logarithmic term in Eq. (5.11) can be neglected since ln(δ/

$\delta_T) \sim 0$ for Pr ~ 1 and $\ln(\delta/\delta_T) \ll P/A$ for $P \gg 1$. Thus, Eq. (5.14) reduces to the following compact form for the momentum/heat transfer analogy (Shenoy, 1992):

$$\left(\frac{2S_{tx}}{C_{fx}}\right) = \left\{1 + \left(\frac{C_{fx}}{2}\right)^{\frac{1}{2}}\left[P - A\sum_{m'=1}^{\infty}\binom{\frac{1}{2}}{m'}\left(\frac{\beta_1^{m'}}{m'}\right)\right]\right\}^{-1} \tag{5.17}$$

A simple integral approach is now followed to get estimates of C_{fx} and β_1 so that the validity of Eq. (5.17) may be substantiated. A usual control volume analysis leads to the momentum balance relation given below:

$$\frac{d}{dx}\int_0^{\delta}\left(u_e u - u^2\right)dy + \frac{du_e}{dx}\int_0^{\delta}(u_e - u)dy = \frac{\tau_w}{\rho} \tag{5.18}$$

For drag reducing fluids which are known to be Newtonian in viscosity but exhibit mild elasticity, one assumes the Fanning friction factor f to be a function of the Reynolds number and the Deborah number. An explicit expression for f covering a range of De between 0 and 10 and that for Reynolds number between 10^4 and 10^6 has been provided (Shenoy, 1988) as given by Eq. (2.24) and reproduced below for convenience:

$$\frac{1}{\sqrt{f}} = 3.57 \log \frac{\mathrm{Re}^{\,1+0.175\mathrm{De}^{(1-0.04\log\mathrm{Re})}}}{(6.5)^{1+0.725\mathrm{De}^{0.7}}} \tag{5.19}$$

A straightforward manipulation can be done to obtain a Blasius-type friction factor Reynolds number relationship as given by Eq. (2.25) and reproduced below for convenience:

$$f = \frac{\alpha}{\mathrm{Re}^{\beta}} \quad \text{for } 5\times10^3 \le \mathrm{Re} \le 10^5 \text{ and } 0 \le \mathrm{De} \le 10 \tag{5.20}$$

α and β are both functions of Deborah number, and their values for varying De are given (Shenoy & Shintre, 1986) and shown in Table 5.1. Equation (5.20) has been used when analyzing turbulent flow in horizontal pipes (Shenoy & Mashelkar, 1983), in curved tubes (Shenoy, Ranade, & Ulbrecht, 1980), in rotating straight tubes (Shenoy, 1986), in annular ducts (Shenoy & Shintre, 1986), and in vertical tubes (Shenoy, 1987).

Following the procedure for Newtonian fluids (Skelland, 1967), a suitable expression for the local skin friction coefficient can be obtained from Eq. (5.20) as follows:

$$C_{fx} = \frac{2\tau_w}{\rho u_e^2} = 2\Omega\left(\frac{\mu}{\rho u_e \delta}\right)^{\beta} \tag{5.21}$$

where

$$\Omega = \frac{\alpha(0.817)^{2-\beta}}{2^{\beta+1}} \tag{5.22}$$

Note that for the Newtonian case

$$\beta = 0.25; \quad \Omega = 0.02332; \quad C_{fx} = 0.04664\left(\frac{\mu}{\rho u_e \delta}\right)^{0.25} \tag{5.23}$$

Equation (5.21) corresponds to the following power-law velocity model for drag reducing fluids (Shenoy, 1988):

$$\frac{u}{u_e} = \left(\frac{y}{\delta}\right)^{\frac{\beta}{2-\beta}} \tag{5.24}$$

Upon substitution of Eqs. (5.21) and (5.24), Eq. (5.18) can be easily solved for δ to give the following:

$$\left(\frac{\delta}{x}\right)\mathrm{Re}_x^{\frac{\beta}{1+\beta}} = \left[\frac{2\Omega(2+\beta)(1+\beta)}{\beta(2-\beta)}\right]^{\frac{1}{1+\beta}} I^{\frac{1}{1+\beta}} \tag{5.25}$$

where

$$I = \frac{\int_0^x u_e^{\frac{3(2+\beta)}{(2-\beta)}}dx}{u_e^{\frac{3(2+\beta)}{(2-\beta)}}x} \tag{5.26}$$

$$\mathrm{Re}_x = \frac{\rho u_e x}{\mu} \tag{5.27}$$

The substitution of Eq. (5.25) into Eqs. (5.21) and (5.14) yields

$$C_{fx}\,\mathrm{Re}_x^{\frac{\beta}{1+\beta}} = \frac{2\Omega}{\left[\frac{2\Omega(2+\beta)(1+\beta)}{\beta(2-\beta)}\right]^{\frac{\beta}{1+\beta}}} I^{\frac{\beta}{1+\beta}} \tag{5.28}$$

and

$$\beta_1 = -\left[\frac{2(2+\beta)(1+\beta)}{\beta(2-\beta)}\right]\overline{m}I \tag{5.29}$$

where

$$\overline{m} = \frac{d\ln u_e}{d\ln x} \tag{5.30}$$

For the special case of $\overline{m}$ being constant, we have the wedge flow for which

$$u_e \propto x^{\overline{m}} \tag{5.31}$$

and

$$I = \left\{1 + \left[\frac{3(2+\beta)}{(2-\beta)}\right]\overline{m}\right\}^{-1} \tag{5.32}$$

The analogy factor based on Eq. (5.17) for the case of the flat plate, i.e., $\overline{m} = 0$ can be written as follows:

$$S_{tx} = \frac{\left(\frac{C_{fx}}{2}\right)}{\left\{1 + \left(\frac{C_{fx}}{2}\right)^{\frac{1}{2}}P\right\}} \tag{5.33}$$

Using Eq. (5.28) for the values of C_{fx}, Eq. (5.33) is plotted in Fig. 5.1 for selected values of Deborah numbers (0, 5, 10, 20) and a typical chosen Reynold number of 10^5 which is a good representative value for drag reducing fluids.

At higher Reynolds numbers, the extent of drag reduction is often larger, but most of the experimental data is normally restricted to a Reynolds number band between 6000 and 2.5×10^5 which is considered to be most convenient for experimentation, as can be seen from the table provided in the review article (Diamant & Poreh, 1976). For drag reducing fluids, the Prandtl number range also is quite restricted, and hence the plot is drawn for the range of Prandtl number between 1 and 100 which is the band of interest. In Fig. 5.1, at De = 0, the obtained curve is no different from that for Newtonian fluids (Nakayama et al., 1984) who have compared it with existing analogies and found good agreement. Under drag reducing conditions, a comparison of the results plotted in Fig. 5.1 would be desirable. However, this cannot be done as there is no theoretical or experimental work available for turbulent flow heat transfer to drag reducing fluids past an isothermal flat plate. Nevertheless, there are some theoretical analyses as well as experimental data for heat transfer in smooth circular pipes during turbulent flow of drag reducing fluids. In order to compare the results of the presented analysis with other existing theoretical expressions, Eq. (5.33) needs to

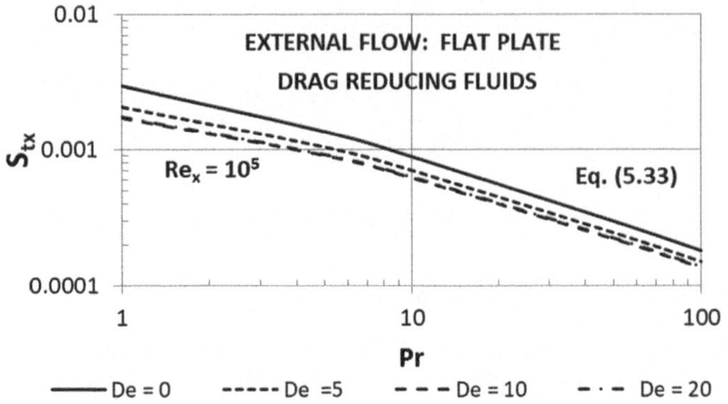

Fig. 5.1 Predictions of the local Stanton number (Eq. (5.33)) for external flow of Newtonian and mildly elastic drag reducing fluids past a flat plate ($\beta_1 = 0$ and $\mathrm{Re}_x = 10^5$) (Shenoy, 1992). (Reprinted with permission from John Wiley and Sons and Copyright Clearance Center through RightsLink)

be adapted from the external flow case to the internal flow situation. Since Eq. (5.33) holds well at the edge of the boundary layer, it is assumed that replacing T_e by T_∞ (ambient temperature of the bulk of the fluid) and replacing u_e by u_m (maximum centerline velocity for pipe flow) retain its validity. u_m is related to $\overline{V}$ (the average velocity) as follows (Shenoy, 1988):

$$u_m = \left(\frac{1}{\overline{\psi}}\right)\overline{V} \tag{5.34}$$

where

$$\overline{\psi} = \frac{(2-\beta)^2}{(4-\beta)} \tag{5.35}$$

Now using the following definitions

$$S_{tx} = \frac{q_w}{\{\rho C_p \overline{V}(T_w - T_\infty)\}} \tag{5.36}$$

and

$$f = \frac{2\tau_w}{\rho \overline{V}^2} \tag{5.37}$$

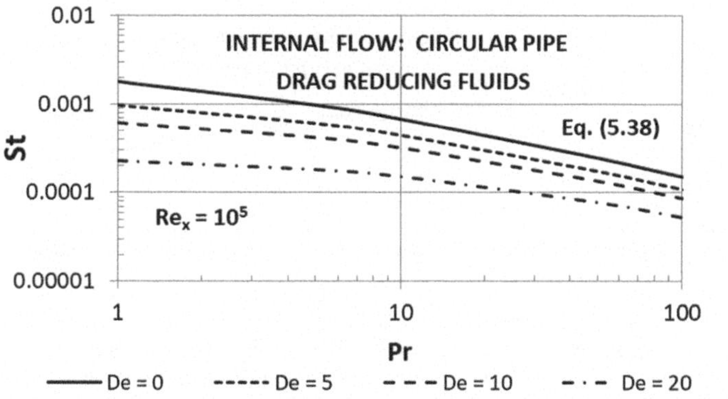

Fig. 5.2 Predictions of the local Stanton number (Eq. (5.38)) for internal flow of Newtonian and mildly elastic drag reducing fluids in smooth circular pipe ($\beta_1 = 0$ and $\mathrm{Re_x} = 10^5$) (Shenoy, 1992). (Reprinted with permission from John Wiley and Sons and Copyright Clearance Center through RightsLink)

The analogy for internal flow can be written as

$$\mathrm{St} = \frac{\left(\frac{f}{2}\right)}{\left\{\frac{1}{\frac{1}{\psi}} + \left(\frac{f}{2}\right)^{\frac{1}{2}}\left[9.24\left(\mathrm{Pr}^{\frac{3}{4}} - 1\right)\right]\right\}} \tag{5.38}$$

where the value for f is used from Eq. (5.20) and a plot is made for varying De (0, 5, 10, 20) and Reynolds number of 10^5 as shown in Fig. 5.2.

To check the propriety of Eq. (5.38), a comparison is made with the following theoretical expressions (Dudukovic, 1988; Kale, 1977) for medium value of De = 5 in Fig. 5.2:

$$\mathrm{St} = \frac{\left(\frac{f}{2}\right)}{\left\{1.2 + \left(\frac{f}{2}\right)^{\frac{1}{2}}(\mathrm{Pr} - 1)b_0\right\}} \tag{5.39}$$

where the equation for b_0 proposed by one researcher (Kale, 1977) is

$$b_0 = \left\{9.2(\mathrm{Pr})^{-0.258} + 1.2(\mathrm{De})(\mathrm{Pr})^{-0.236}\right\} \tag{5.40}$$

where the equation for b_0 proposed by another researcher (Dudukovic, 1988) is

$$b_0 = \left\{9.2(\mathrm{Pr})^{-0.255} + 1.2(\mathrm{De})(\mathrm{Pr})^{-0.255} - 0.204(\mathrm{De})^2(\mathrm{Pr})^{-0.255}\right\} \tag{5.41}$$

It should be noted that the first two terms in Eq. (5.41) are almost identical to those proposed in Eq. (5.40). In that sense, Eq. (5.41) provides an extra correction

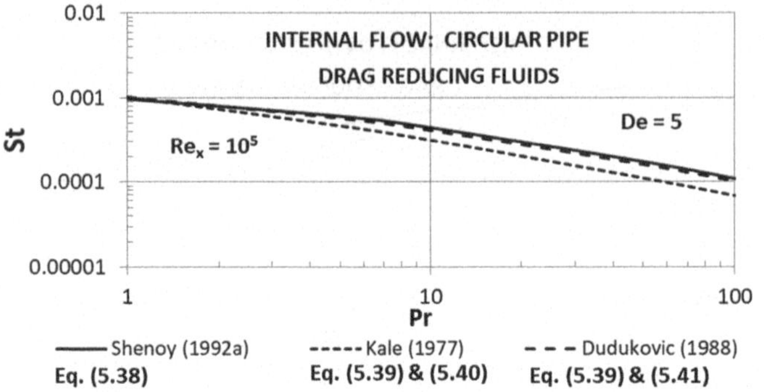

Fig. 5.3 Comparison of the predictions of the local Stanton number from Shenoy (1992), Kale (1977), and Dudukovic (1988) for internal flow of mildly elastic drag reducing fluids in a smooth circular pipe ($\beta_1 = 0$, De $= 5$ and Re$_x = 10^5$) (Shenoy, 1992). (Reprinted with permission from John Wiley and Sons and Copyright Clearance Center through RightsLink)

term by way of a higher-order power of De. It is seen from Fig. 5.3 that Eq. (5.38) gives a very close match to Eq. (5.41) but deviates slightly from Eq. (5.40). It is unfortunate that Eq. (5.41) cannot be compared at higher values of De especially for the maximum drag reducing case. This happens because Eq. (5.41) has a negative sign for the third term and hence tends to give erroneous results at values of De $\geq$ 5.88. Equation (5.40) predicts a much higher reduction in Stanton number as shown in Fig. 5.3 and does so more for the maximum drag reducing case. Predictions from Eq. (5.40) have been shown (Kale, 1977) to match reasonably well with experimental data. A plot of the kind (Nu$_{\text{experimental}}$ vs Nu$_{\text{predicted}}$) if done using Eq. (5.38) would also show good agreement. It should be noted that for De $= 0$, all three equations give almost identical results. Whereas most would use $1/\overline{\psi}$ as equal to 1.2, Eq. (5.38) uses Eq. (5.35) to determine this value for different Deborah numbers. In fact, at De $= 0$, Eq. (5.38) predicts $1/\overline{\psi} = 1.22$. Further, it can be seen that the (Shenoy, 1992) model presented above predicts St $\propto$ Pr$^{-3/4}$ at higher Prandtl numbers for all values of De. This trend has been noted (Diamant & Poreh, 1976) as the preferred trend based on earlier theoretical analysis and experimental data. It is thus reasonable to expect that Eqs. (5.17), (5.33), and (5.38) would provide results with a good deal of accuracy for the entire range of Deborah numbers including the case where Eq. (5.41) fails and where Eq. (5.40) seems to overpredict. It is worth mentioning at this point that the entire analysis was based on a modification of the velocity profile to take care of the viscoelasticity of the drag reducing polymer. There is experimental evidence that the velocity profile alone does not explain the measurements of Reynolds stresses in dilute drag reducing polymer solution flows through pipes and channels (Bewersdorff, 1984; Patterson, Chosnek, & Zakin, 1977; Thielen, 1981; Willmarth, Wei, & Lee, 1987). It was found that the sum of the Reynolds stress and the conventional molecular stress is only two-thirds of the force produced by the streamwise pressure gradient. This would mean that a viscoelastic term may be needed to be added to the shear stress in addition to the modification of the velocity profile. This was not done (Shenoy, 1992) to maintain the simplicity of the analysis. However, any extensions of this work may warrant the inclusion of the term in case it is found that the theory does not match the experiments as closely as expected.

Vertical Tubes

For mildly elastic drag reducing fluids, the reduction in drag is accompanied by a larger reduction in the forced convection heat transfer rate (Astarita & Marrucci, 1966; Gupta et al., 1967; Wells, 1968), and hence drag reducing fluids have had limited use in heat transfer circumstances. However, in any forced convection heat transfer situation, density differences are bound to arise, and the effects of natural convection then cannot be ignored. Often, buoyancy effects take a high enough magnitude, and hence it is the combined effect of the forced and free convection that truly determines the heat transfer rate. Mixed convection heat transfer to non-Newtonian fluids is more difficult to analyze due to the complex equations involved in describing such fluids (Shenoy, 1988). However, due to the obvious realization of the importance of such studies, there have been attempts to investigate the combined effect of forced and free convection in power-law fluids as well as viscoelastic fluids in both internal (De Young & Scheele, 1970; Marner & McMillan, 1972; Marner & Rehfuss, 1972; Metzner & Gluck, 1960; Oliver & Jenson, 1964; Scheele & Greene, 1971;Shenoy, 1984a, 1987) and external (Shenoy, 1980a, 1980b, 1984b) flow situations. But all those efforts have been directed to the laminar flow situation except for the ones (Shenoy, 1984a, 1987) that give a theoretical analysis of the mixed convection problem to turbulent flowing inelastic power-law fluids and mildly elastic drag reducing fluids, respectively.

The present section puts forth an approximate theoretical analysis (Shenoy, 1987) of the effect of buoyancy on the heat transfer to drag reducing fluids for upward flow in vertical pipes under turbulent conditions. A criterion is suggested for controlling the reduction in heat transfer due to natural convection to less than 5%.

The system under consideration is a drag reducing fluid with a fully developed velocity profile and at a uniform temperature T_i, flowing under turbulent forced convection conditions in a vertical, circular tube of radius R_t and maintained at a constant temperature T_w. It is obvious that at a high enough temperature difference between the tube wall and the fluid, buoyancy forces would begin to act due to reduced fluid density and tend to reduce the shear stress at a distance from the surface as given by the following integral:

$$\Delta\tau = \int_0^{\delta_B} (\rho_b - \rho) g dy \tag{5.42}$$

A consequence of this shear stress decrease is that the turbulence production is reduced and the structure of the turbulent flow becomes more like that of a buoyancy-free flow at a reduced Reynolds number. If it is assumed that the density variation with temperature is gradual, then the buoyant layer and the thermal layer could be taken to be identical in thickness and the temperature gradient in the thermal layer approximately written as $(T_w - T_\infty)/\delta_B$. Now if an integrated density $\overline{\rho}$ of the following form is defined

$$\overline{\rho} = \frac{1}{T_w - T_\infty}\int_{T_\infty}^{T_w} \rho\, dT \tag{5.43}$$

then Eq. (5.42) can be written as

$$\Delta\tau = \delta_B g(\rho_b - \rho) \tag{5.44}$$

Knowing the approximate relationship between the thermal and momentum boundary-layer thickness, the following equation can be written:

$$\frac{\delta_T}{\delta_M} = \frac{\delta_B}{\delta_M}\frac{1}{\mathrm{Pr}^{\frac{1}{2}}} \tag{5.45}$$

where Pr is the Prandtl number equal to $Cp\mu_b/k$.

A dimensionless wall-layer thickness δ_M^+ is now defined as follows:

$$\delta_M^+ = \frac{(\tau_w \rho_w)^{\frac{1}{2}}\delta_M}{\mu_w} \tag{5.46}$$

and Eq. (5.44) is rewritten using Eqs. (5.45) and (5.46) as

$$\Delta\tau = \frac{\delta_M^+ \mu_w g(\rho_b - \overline{\rho})}{(\tau_w \rho_w \mathrm{Pr})^{\frac{1}{2}}} \tag{5.47}$$

δ_M^+ denotes the dimensionless thickness of the combined sublayer and the buffer layer in the above equation. To determine its value, the distance of the edge of the buffer layer from the pipe wall needs to be known. This could be found by evaluating the intersection of the equations for the velocity profile in the buffer layer and the turbulent core. The velocity profile for the turbulent core in smooth circular pipes is given (Seyer & Metzner, 1969) in Eq. (2.7) as

$$u^+ = 2.46\ \ln y^+ + 5.6 + 1.55\ \mathrm{De} \quad (\text{for } 0 \le \mathrm{De} \le 10) \tag{5.48}$$

The expression for the velocity profile in the buffer layer can be taken from the elastic sublayer concept (Virk, 1971, 1975) which is characteristic of drag reduction. He proposed that for drag reducing fluids, there existed an elastic sublayer which grew with increasing drag reduction until a maximum asymptotic value. The velocity profile in the elastic sublayer segment was given (Virk, 1975) as

$$u^{+} = 11.7\ \ln y^{+} - 17.0 \tag{5.49}$$

To determine the extent of the elastic sublayer, an intersection of Eqs. (5.48) and (5.49) is taken giving

$$\delta_{\mathrm{M}}^{+} = e^{2.45+0.17\mathrm{De}} \tag{5.50}$$

Combining Eqs. (5.47) and (5.50) gives the expression for the fractional reduction in shear stress due to buoyancy as

$$\frac{\Delta\tau}{\tau_{\mathrm{w}}} = \frac{e^{2.45+0.17\mathrm{De}}\mu_{\mathrm{w}}g(\rho_{\mathrm{b}} - \overline{\rho})}{\tau_{\mathrm{w}}^{\frac{3}{2}}\rho_{\mathrm{w}}^{\frac{1}{2}}\mathrm{Pr}^{\frac{1}{2}}} \tag{5.51}$$

This can be rewritten in the following form as

$$\frac{\Delta\tau}{\tau_{\mathrm{w}}} = \frac{2\sqrt{2}e^{2.45+0.17\mathrm{De}}\mathrm{Gr}}{f^{\frac{3}{2}}\mathrm{Pr}^{\frac{1}{2}}\,\mathrm{Re}^{\,3}}\left(\frac{\mu_{\mathrm{w}}}{\mu_{\mathrm{b}}}\right)\left(\frac{\rho_{\mathrm{b}}}{\rho_{\mathrm{w}}}\right)^{\frac{1}{2}} \tag{5.52}$$

where

$$\mathrm{Gr} = \frac{\rho_{\mathrm{b}}(\rho_{\mathrm{b}} - \overline{\rho})gD^{3}}{\mu_{\mathrm{b}}^{2}} \tag{5.53}$$

$$\mathrm{Re} = \frac{\rho_{\mathrm{b}}\overline{U}D}{\mu_{\mathrm{b}}} \tag{5.54}$$

$$f = \frac{2\tau_{\mathrm{w}}}{\rho_{\mathrm{b}}\overline{U}^{2}} \tag{5.55}$$

Using the Blasius type of expression given (Shenoy & Mashelkar, 1983) in Eq. (2.25) for the friction factor versus Reynolds number, Eq. (5.52) is rewritten as

$$\frac{\Delta\tau}{\tau_{\mathrm{w}}} = \left(\frac{2}{\alpha}\right)^{\frac{3}{2}}\left[e^{2.45+0.17\mathrm{De}}\right]\frac{\mathrm{Gr}}{\mathrm{Pr}^{\frac{1}{2}}\,\mathrm{Re}^{\frac{3(2-\beta)}{2}}}\left(\frac{\mu_{\mathrm{w}}}{\mu_{\mathrm{b}}}\right)\left(\frac{\rho_{\mathrm{b}}}{\rho_{\mathrm{w}}}\right)^{\frac{1}{2}} \tag{5.56}$$

It can thus be seen that the Grashof number and the Reynolds number combined in the form $\mathrm{Gr}/\mathrm{Re}^{\frac{3(2-\beta)}{2}}$ truly control the extent to which buoyancy forces would modify the shear stresses. Based on the assumption made earlier, that the modified flow is like a buoyancy-free flow at some reduced value of Reynolds number Re', a relationship between the nominal and the reduced shear stress with the respective Reynolds numbers can be written as

$$\frac{\tau'_{\mathrm{w}}}{\tau_{\mathrm{w}}} = \left(\frac{f'}{f}\right)\left(\frac{\mathrm{Re}'}{\mathrm{Re}}\right)^2 \tag{5.57}$$

Again, using the Blasius type of expression given (Shenoy & Mashelkar, 1983) in Eq. (2.25) for the friction factor versus Reynolds number, Eq. (5.57) is modified to give

$$\frac{\tau'_{\mathrm{w}}}{\tau_{\mathrm{w}}} = \left(\frac{\mathrm{Re}'}{\mathrm{Re}}\right)^{2-\beta} \tag{5.58}$$

The established empirical form relating Nusselt number to Reynolds number for Newtonian fluids, namely, $\mathrm{Nu} \propto \mathrm{Re}^{0.8}$, is used as a first approximation to give

$$\frac{\mathrm{Nu}'}{\mathrm{Nu}} = \left(\frac{\tau'_{\mathrm{w}}}{\tau_{\mathrm{w}}}\right)^{\frac{0.8}{2-\beta}} \tag{5.59}$$

Noting that $\Delta\tau = \tau_{\mathrm{w}} - \tau'_{\mathrm{w}}$, the expression for reduced Nusselt number can be written as

$$\frac{\mathrm{Nu}'}{\mathrm{Nu}} = \left\{1 - \left(\frac{2}{\alpha}\right)^{\frac{3}{2}}\left[e^{2.45+0.17\mathrm{De}}\right]\frac{\mathrm{Gr}}{\mathrm{Pr}^{\frac{1}{2}}\,\mathrm{Re}^{\frac{3(2-\beta)}{2}}}\left(\frac{\mu_{\mathrm{w}}}{\mu_{\mathrm{b}}}\right)\left(\frac{\rho_{\mathrm{b}}}{\rho_{\mathrm{w}}}\right)^{\frac{1}{2}}\right\}^{\frac{0.8}{2-\beta}} \tag{5.60}$$

Using appropriate values of β from Table 5.2 in Eq. (5.59) shows that a 10% reduction in shear stress induced by buoyancy would lead to a reduction in heat transfer coefficient for Newtonian fluids of 4.7%, for a moderate drag reducing fluid ($\mathrm{De} = 10$) of 4.9%, and for the maximum drag reduction ($\mathrm{De} \geq 20$) of 5.6%.

Now using Eq. (5.60), a criterion for reduction of 5% or less in the heat transfer coefficient can be set up as follows:

$$\frac{\mathrm{Gr}}{\mathrm{Pr}^{\frac{1}{2}}\,\mathrm{Re}^{\frac{3(2-\beta)}{2}}}\left(\frac{\mu_{\mathrm{w}}}{\mu_{\mathrm{b}}}\right)\left(\frac{\rho_{\mathrm{b}}}{\rho_{\mathrm{w}}}\right)^{\frac{1}{2}} < C \tag{5.61}$$

where

$$C = 0.1\left(\frac{\alpha}{2}\right)^{\frac{3}{2}}\left[e^{-(2.45+0.17\mathrm{De})}\right] \tag{5.62}$$

The values of C for varying Deborah numbers are given in Table 5.2. It is clear from the table that with increasing drag reduction, it would become increasingly

Table 5.2 Values of C for different values of De, α, and β (Shenoy, 1987)

De	α	β	C
0	0.079	0.250	6.77×10^{-5}
1	0.0782	0.262	5.63×10^{-5}
2	0.0787	0.271	4.61×10^{-5}
3	0.0741	0.278	3.70×10^{-5}
4	0.0726	0.285	3.01×10^{-5}
5	0.0689	0.289	2.36×10^{-5}
6	0.0655	0.292	1.84×10^{-5}
7	0.0662	0.301	1.58×10^{-5}
8	0.0687	0.312	1.41×10^{-5}
9	0.0732	0.324	1.31×10^{-5}
10	0.0762	0.334	1.17×10^{-5}
≥ 20	0.420	0.550	–

difficult to maintain the reduction in heat transfer due to the buoyancy effect to less than 5%. Previous studies on heat transfer to drag reducing fluids (Astarita & Marrucci, 1966; Gupta et al., 1967; Mizushina, Usui, & Yamamoto, 1975; Wells, 1968) have indicated that the reduction in heat transfer is far more conspicuous than the reduction in momentum transfer. The analysis presented in this section, in fact, shows that, at least in vertical tubes, natural convection would further reduce the heat transfer coefficient in the case of turbulent flow of drag reducing fluids. This would happen because the reduction in shear stress due to the drag reducing additive along with that due to the buoyancy forces would have a cumulative effect and hence result in a very large reduction in the heat transfer coefficient. Thus, what is gained through a reduction in drag is more than lost due to a reduction in heat transfer especially in the presence of buoyancy forces.

References

Astarita, G., & Marrucci, G. (1966). Heat transfer in viscoelastic liquids in turbulent flow. In *Proc. 36th Int. Cong. Ind. Chem., Brussels*.

Bewersdorff, H. W. (1984). Effect of injected polymer thread on turbulent properties of pipe flow. In R. H. J. Sellin & R. T. Moses (Eds.), *Proc. third intl. conf. on drag reduction* (pp. B4.1–B4.8).

Cho, Y. I., & Hartnett, J. P. (1980). Analogy for viscoelastic fluids – momentum, heat and mass transfer in turbulent pipe flow. *Letters in Heat and Mass Transfer, 7*, 339–346.

De Young, S. H., & Scheele, G. F. (1970). Natural convection distorted non-newtonian flow in a vertical pipe. *AICHE Journal, 16*(5), 712–717.

Diamant, Y., & Poreh, M. (1976). Heat transfer in flows with drag reduction. *Advances in Heat Transfer, 12*, 77–113.

Dudukovic, A. (1988). Encyclopedia of fluid mechanics, vol. 7, ch. 12. In N. P. Cheremisinoff (Ed.), *Analogies between momentum, heat and mass transfer in dilute polymer solutions* (pp. 341–357). Houston, TX: Gulf Publishing Co.

Friend, W. L., & Metzner, A. B. (1958). Turbulent heat transfer inside tubes and the analogy among heat, mass, and momentum transfer. *AICHE Journal, 4*(4), 393–402.

Gasljevic, K., Aguilar, G., & Matthys, E. F. (2000). Buoyancy effects on heat transfer and temperature profiles in horizontal pipe flow of drag-reducing fluids. *International Journal of Heat and Mass Transfer, 43*(23), 4267–4274.

Gupta, M. K., Metzner, A. B., & Hartnett, J. P. (1967). Turbulent heat transfer characteristics of viscoelastic fluids. *International Journal of Heat and Mass Transfer, 10*, 1121–1224.

Jayatillaka, C. I. (1969). The influence of prandtl number and surface roughness on the resistance of the laminar sublayer to momentum and heat transfer. In U. Grigull & E. Hahne (Eds.), *Progress in heat and mass transfer* (pp. 193–329). New York: Pergamon Press.

Kale, D. D. (1977). An analysis of heat transfer to turbulent flow of drag reducing fluids. *International Journal of Heat and Mass Transfer, 20*(10), 1077–1081.

Kawase, Y., & Ulbrecht, J. (1982). Turbulent heat and mass transfer to turbulent flow of drag reducing fluids. *Chemical Engineering Science, 37*(7), 1039–1046.

Marner, W. J., & McMillan, H. K. (1972). Combined free and forced laminar non-newtonian convection in a vertical tube with constant wall temperature. *Chemical Engineering Science, 27* (3), 473–488.

Marner, W. J., & Rehfuss, R. A. (1972). Buoyancy effects on fully developed laminar non-newtonian flow in a vertical tube. *Chemical Engineering Journal, 3*, 294–300.

Metzner, A. B., & Friend, P. S. (1959). Heat transfer to turbulent non-newtonian fluids. *Industrial and Engineering Chemistry, 51*, 879–882.

Metzner, A. B., & Gluck, D. F. (1960). Heat transfer to non-newtonian fluids under laminar flow conditions. *Chemical Engineering Science, 12*, 185–190.

Mizushina, T., Usui, H., & Yamamoto, T. (1975). Turbulent heat transfer to viscoelastic fluids flow in pipe. *Letters in Heat Mass Transfer, 2*(1), 19–26.

Nakayama, A., Koyama, H., & Ohsawa, S. (1984). Momentum/heat transfer analogy for turbulent boundary layers in mild pressure gradients. *AIAA Journal, 22*, 841–844.

Oliver, D. R., & Jenson, V. G. (1964). Heat transfer to pseudoplastic fluids in laminar flow in horizontal tubes. *Chemical Engineering Science, 19*(2), 115–129.

Patterson, G. K., Chosnek, J., & Zakin, J. L. (1977). Turbulence structure in drag reducing polymer solutions. *Physics of Fluids, 20*(10), S89–S99.

Poreh, M., & Paz, U. (1968). Turbulent heat transfer to dilute polymer solutions. *International Journal of Heat and Mass Transfer, 11*(5), 805–818.

Scheele, G. F., & Greene, H. L. (1971). Non-newtonian flow stability in a heated pipe at low reynolds numbers. *Industrial and Engineering Chemistry Fundamentals, 10*(1), 102–113.

Seyer, F. A., & Metzner, A. B. (1969). Turbulence phenomena in drag reducing systems. *AICHE Journal, 15*(3), 426–434.

Shenoy, A. V. (1980a). A correlating equation for combined laminar forced and free convection heat transfer to power-law fluids. *AICHE Journal, 26*(3), 505–507.

Shenoy, A. V. (1980b). Combined laminar forced and free convection heat transfer to viscoelastic fluids. *AICHE Journal, 26*(4), 683–685.

Shenoy, A. V. (1984a). Natural convection effects o heat transfer to power-law fluids flowing under turbulent conditions in vertical pipes. *International Communications in Heat Mass Transfer, 11* (5), 467–476.

Shenoy, A. V. (1984b). Laminar mixed convection heat transfer from an isothermal inclined flat plate to power-law fluids. *AICHE Journal, 30*(5), 824–826.

Shenoy, A. V. (1986). Turbulent flow of mildly elastic fluids through rotating straight circular tubes. *Journal of Applied Sciences Research, 43*(1), 39–54.

Shenoy, A. V. (1987). Effects of bouyancy on heat transfer during turbulent flow of drag reducing fluids in vertical pipes. *Warme- und Stoffubertragung, 21*(1), 15–18.

Shenoy, A. V. (1988). Encyclopedia of fluid mechanics, vol. 7, ch. 16. In N. P. Cheremisinoff (Ed.), *Turbulent flow velocity profiles in drag-reducing fluids* (pp. 479–503). Houston, TX: Gulf Publishing Co.

Shenoy, A. V. (1992). Momentum/heat transfer analogy for drag reducing fluids during turbulent boundary layer flow with small pressure gradients. *Canadian Journal of Chemical Engineering, 70*(2), 375–380.

Shenoy, A. V., & Mashelkar, R. A. (1983). Engineering estimate of hydrodynamic entrance lengths in non-newtonian turbulent flow. *Industrial and Engineering Chemistry Process Design and Development, 22*(1), 165–168.

Shenoy, A. V., & Shintre, S. N. (1986). Developing and fully developed turbulent flow of drag reducing fluids in an annular duct. *The Canadian Journal of Chemical Engineering, 64*(2), 190–195.

Shenoy, A. V., & Talathi, M. M. (1985). Turbulent pipe flow velocity profile model for drag-reducing fluids. *AICHE Journal, 31*(3), 520–522.

Shenoy, A. V., Ranade, V. R., & Ulbrecht, J. J. (1980). Turbulent flow of mildly viscoelastic liquids in curved tubes. *Chemical Engineering Communications, 5*(5–6), 269–286.

Skelland, A. H. (1967). *Non-newtonian flow and heat transfer*. New York: Wiley.

Smith, K. A., Keuroghlian, P. S., Virk, P. S., & Merrill, E. W. (1969). Heat transfer to drag reducing polymer solurions. *AICHE Journal, 15*(2), 294–297.

Thielen, W. (1981). Turbulenzstruktur in der rohrestromung viscoelastischer fluussigkeiten, dissertation, RWTH-Aachen, F. R. G.; partly communicated by P. Schummer and W. Thielen, structure of turbulence in viscoelastic fluids. *Chemical Engineering Communications, 4*(4–5), 593–606. (1980).

Virk, P. S. (1971). Drag reduction in rough pipes. *Journal of Fluid Mechanics, 45*(2), 225–246.

Virk, P. S. (1975). Drag reduction fundamentals. *AICHE Journal, 21*(4), 625–656.

Wells, C. S. (1968). Turbulent heat transfer in drag reducing fluids. *AICHE Journal, 14*(3), 406–410.

Willmarth, W. W., Wei, T., & Lee, C. O. (1987). Laser anemometer measurements of Reynolds stress in a turbulent channel flow with drag reducing polymer additives. *Physics of Fluids, 30*(4), 933–935.

Chapter 6
Natural, Forced, and Mixed Convection Heat Transfer in External Flows Through Porous Media

Vertical Flat Plate

The interest in heat transfer through porous media is mainly due to several practical applications of the subject matter in geothermal engineering, thermal insulation systems, ceramic processing, enhanced oil recovery, filtration processes, chromatography, etc. The tremendous research work that has gone into the area of convective heat transfer in porous media can be adjudged from the many reviews that have appeared on this subject (Bejan, 1987; Cheng, 1978; Combarnous & Bories, 1975; Kafoussias, 1990; Shenoy, 1994; Tien & Vafai, 1990; Trevisan & Bejan, 1990). Bulk of the literature focuses on steady-state heat transfer in porous media to Newtonian fluids, and there were no studies until the late 1980s wherein non-Newtonian fluid-saturated porous media were considered as can be seen by their absence in the comprehensive review articles (Shenoy, 1986, 1988).

This area began to attract attention (Cao & Cui, 2015; Chen & Chen, 1988a, 1988b; Nakayama & Koyama, 1991; Nakayama & Shenoy, 1993; Pascal, 1990a, 1990b; Pascal & Pascal, 1989; Wang & Tu, 1989; Wang, Tu, & Zhang, 1990) because of the realization that a number of fluids which come in contact with porous media exhibit non-Newtonian flow behavior, especially in ceramic processing, enhanced oil recovery, and filtration.

The initial simplest free convection flow of non-Newtonian fluids past an isothermal vertical flat plate embedded in a porous medium (Chen & Chen, 1988a) was later extended (Chen & Chen, 1988b) to include other body shapes such as horizontal cylinders and spheres. Boundary-layer flow and heat transfer to non-Newtonian fluids in porous media were explored (Wang & Tu, 1989), and expressions for the local Nusselt numbers for the forced and free convection of a Herschel-Bulkley type non-Newtonian fluid past an isothermal semi-infinite plate in porous media were obtained. Nonlinear effects of non-Newtonian fluids on natural convection in a porous medium were analyzed (Pascal & Pascal, 1989) for the two cases of constant temperature and constant heat flux boundary-layer flow of

A. Shenoy, *Rheology of Drag Reducing Fluids*,
https://doi.org/10.1007/978-3-030-40045-3_6

Herschel-Bulkley fluid along a heated vertical cylinder. The rheological effects of some non-isothermal flows through a porous medium were also studied (Pascal, 1990a). An unsteady-state solution for the case of a shear-thinning fluid in the presence of a yield stress was obtained. The case of two-phase flows of non-Newtonian fluids through a porous medium was also analyzed (Pascal, 1990b). Buoyancy-induced flow of non-Newtonian fluids over a non-isothermal body of arbitrary shape in a fluid-saturated porous medium was treated (Nakayama & Koyama, 1991). Combined free and forced convection heat transfer to power-law fluid-saturated porous media was analyzed (Nakayama & Shenoy, 1993), and possible similarity solutions for vertical flat plates, cones, horizontal cylinders, and spheres were presented. A numerical study of natural convection during heat and mass transfer along an infinite vertical plate in porous media was done (Cao & Cui, 2015) using the similarity transformation and shooting method. The effects of power-law viscosity on temperature and mass were studied for the case of both the thermal diffusivity and mass diffusivity varying as a power-law function. The effects of some physical quantities such as the power-law fluids' index, the threshold pressure gradient parameter, and the buoyancy ratio on heat and mass transfer characteristics were discussed.

The one common feature of all the abovementioned non-Newtonian fluid-saturated porous media studies is that they only dealt with the Darcy flow model, which truly neglects the boundary and inertia effects on fluid flow and heat transfer. It is well-known that when the velocity increases, the flow enters a nonlinear laminar regime and then the porous inertia effects are no longer negligible. For Newtonian fluid flow, a square velocity term in addition to the Darcian velocity term was proposed (Forchheimer, 1901) to account for this effect and called the Forchheimer term (Muskat, 1946). This pioneering work was followed by other proposals (a) for mathematically describing non-Darcy flows (Ergun, 1952; Ward, 1964) and (b) for emphasizing its importance in different flow and heat transfer situations (Bejan & Poulikakos, 1984; Chamkha & Ben-Nakhi, 2007; Fand, Steinberger, & Cheng, 1986; Ingham, 1986; Nield & Joseph, 1985; Plumb & Huenefeld, 1981; Poulikakos & Bejan, 1985; Vasantha, Pop, & Nath, 1986). When dealing with non-Darcy flow, thermal boundary-layer effects (Brinkman, 1947) also need to be considered by the inclusion of the Brinkman term, and there are studies (Lauriat & Prasad, 1987; Tong & Subramanian, 1985) which take this aspect into account for Newtonian fluids.

Forchheimer effect would certainly be relevant in non-Newtonian inelastic power-law fluids, especially in ceramic processing and enhanced oil recovery, wherein low porosity densely packed media are encountered. Practical situations involving high porosity flow of power-law fluids are rather hard to find, and, therefore, the Brinkman effect would not be as important when dealing with power-law fluids as the Forchheimer effect. For non-Newtonian fluids, the problem of steady-state Darcy-Forchheimer forced, natural, and mixed convection in power-law fluid-saturated porous medium has been studied comprehensively (Shenoy, 1993) for all possible cases. Similarity solutions were found for the case of the isothermal vertical flat plate, and explicit heat transfer expressions have been proposed which give predictions within 2.5% of the exact solutions. A unified transformation from which all possible similarity solutions could be deduced for Darcy and non-Darcy convective flow of power-law fluids has been proposed (Nakayama & Shenoy, 1992). The coupled steady, laminar, and heat and mass transfer by mixed convection was considered (Chamkha & Ben-Nakhi, 2007) from a semi-infinite, isothermal, vertical permeable surface embedded in a non-Darcian porous medium for a non-Newtonian power-law fluid in the presence of temperature-dependent heat generation or absorption. A single mixed convection parameter for the entire range of free-forced-mixed convection was used, and the obtained non-similar differential equations were solved numerically by the implicit finite-difference method. The effect of concentration to thermal buoyancy ratio, power-law fluid viscosity index, heat generation and absorption parameter, the Lewis number, and the porous medium inertia parameter on the local Nusselt and the Sherwood numbers was demonstrated.

All the abovementioned non-Newtonian fluid-saturated porous media studies have dealt with only inelastic non-Newtonian fluids. None have considered the fluid to be elastic though in non-Newtonian fluid mechanics, the effects of elasticity are of considerable interest, but this aspect has not been studied much in convective heat transfer in porous medium because of the lack of knowledge about the modification required in the governing equation to include the elastic effects. Starting from first principles, such a modification of the Darcy's law was derived (Shenoy, 1992). The type of viscoelastic fluids considered are elastic fluids with constant viscosity (Boger, 1977/78; Choplin, Carreau, & Aitkadi, 1983). This helps in bringing out the effect of elasticity exclusively. If viscoelastic fluids, which are elastic and shear-thinning are considered, then one property overshadows the other and the effects of elasticity and shear-thinning cannot be separated out. In the present section, the analysis (Shenoy, 1992) of steady-state Darcy natural, forced, and mixed convection past an isothermal vertical flat plate embedded in a viscoelastic fluid-saturated porous medium is presented.

Appropriate transformations of variables are done to obtain similarity solutions from the governing equations. The results of the analysis are compared with existing exact and approximate solutions in the limiting cases to show an agreement within 2.5% of reported exact values.

The porous media is assumed to be composed of individual discrete particles and completely filled, i.e., saturated with an elastic fluid of constant viscosity but no shear-thinning characteristics. Steady flow of the fluid is assumed to take place through a cylindrical element as shown in Fig. 6.1 which contains sufficient number of particles such that the porosity of the element is representative of the entire porous medium.

Summation of the forces acting in the x-direction on the fluid in the element results in a zero net force under the assumed steady-state conditions and can be written in the simplified form as given by Eq. (6.1) and shown below for convenience:

Fig. 6.1 Forces acting on fluid within an imaginary cross section of a cylindrical porous medium element (Rumer, 1969). (Reprinted with permission from Oxford University Press)

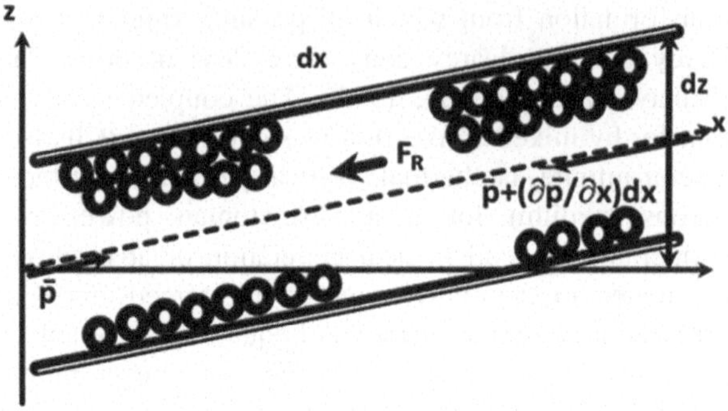

$$-\frac{\partial p}{\partial x}-\rho g_x-\frac{F_R}{\epsilon dA_c dx} \tag{6.1}$$

where p is the pressure, ρ the fluid density, g_x is the component of the gravitational force in the x-direction, ϵ is the porosity of the element, $\epsilon dA_c dx$ is the fluid volume of the element, dA_c is the total cross-sectional area of the element, and F_R is the total frictional resistance offered by all the N particles in the element. Thus,

$$F_R = Nf_p \tag{6.2}$$

where f_p is the frictional resistance or drag on a single particle and N is the total number of particles given as

$$N=\frac{(1-\epsilon)dA_c dx}{\beta'' d_p^3} \tag{6.3}$$

where d_p is particle diameter and β'' is the shape factor which takes a value of $\pi/6$ for spheres.

The general expression for frictional resistance of a single particle can be written as

$$f_p = C_D\alpha'' d_p^2\left(\frac{\rho u_s^2}{2}\right) \tag{6.4}$$

where C_D is the drag coefficient. The influence of elasticity on the drag force of a sphere in the creeping flow regime has been studied by several workers and has been described (Chhabra, 1986) in detail. The theoretical analysis of slow flow of viscoelastic fluids past a sphere (Caswell & Schwarz, 1962; Giesekus, 1963; Leslie & Tanner, 1961; Mena & Caswell, 1974) is based on a perturbation analysis of the Newtonian solution and predicts a departure from the Stokes' law that may be expressed in a general form to any shaped particle as

$$C_D = \left(\frac{2\,\epsilon\lambda''}{\alpha''\,\text{Re}}\right)\left[1 - a_0''\left(\frac{\lambda u}{R_p}\right)^2 + O\left(\frac{u}{R_p}\right)^4\right]$$
$$\text{for We} = \left(\frac{\lambda u}{R_p}\right) \ll 1 \text{ and Re} = \left(\frac{\rho u d_p}{\mu_0}\right) \ll 1 \tag{6.5}$$

where a_0'' is a constant, λ is the fluid relaxation time, R_p is the particle radius, and We is the Weissenberg number that is commonly used to give a measure of fluid elasticity (Metzner, White, & Denn, 1966; White, 1964). λ'' is a coefficient dependent upon particle shape and size and which reaches a limiting minimum value of 3π for a single sphere, Re is a conventional Reynolds number for a constant viscosity fluid, and u is the Darcian or superficial velocity defined as

$$u = \epsilon u_s \tag{6.6}$$

Experimental evidence (Broadbent & Mena, 1974) confirms that, for elastico-viscous fluids, the departure from the Newtonian drag is a quadratic function of velocity as given in Eq. (6.5). The constant term α'' can be obtained from the first-order approximation of the theoretical work (Giesekus, 1963) or experimental data (Broadbent & Mena, 1974), and Eq. (6.5) can be written approximately as

$$C_D = \left(\frac{2\,\epsilon\lambda''}{\alpha''\,\text{Re}}\right)\left[1 - 0.21\left(\frac{\lambda u}{R_p}\right)^2\right] \quad \text{for We} \ll 1 \text{ and Re} \ll 1 \tag{6.7}$$

The above equation has rather severe restrictions on the Weissenberg number and is valid only for slight degree of elasticity. Experimental data (Chhabra, Uhlherr, & Boger, 1980) has shown that the drag coefficient departs considerably from the Newtonian value at We > 0.1 and reaches an asymptotic reduction of 26% for We = 0.7. Hence it is unreasonable to expect that the reduction in drag coefficient is of the order of We^2. In fact, correlating the experimental data (Chhabra et al., 1980) with an equation of the form given by Eq. (6.5) showed a poor fit. However, assuming that the dependence of drag would be of the order of We rather than We^2, the following equation was found to correlate the experimental data (Chhabra et al., 1980) very well.

$$C_D = \left(\frac{2\,\epsilon\lambda''}{\alpha''\,\text{Re}}\right)\left[1 - 0.37\left(\frac{\lambda u}{R_p}\right)\right] \quad \text{for We} \le 0.7 \tag{6.8}$$

and

$$C_D = \left(\frac{2\,\epsilon\lambda''}{\alpha''\,\text{Re}}\right)[0.74] \quad \text{for We} \ge 0.7 \tag{6.9}$$

In the subsequent treatment, Eqs. (6.8) and (6.9) will be used as they are the outcome of controlled experiments with fluids which exhibit elasticity but no significant shear-thinning characteristics. If an equation based on the theoretical analysis is used, then the applicability range of the equation gets restricted to values of We $\ll 1$, whereas Eq. (6.9) is valid even up to We $= 2$. This would allow the true influence of isolated elasticity to be brought out. Combining Eqs. (6.1), (6.2), (6.3), (6.4), (6.5), and (6.6) gives

$$-\frac{\partial p}{\partial x} - \rho g_{\mathrm{x}} - \frac{\alpha''(1-\epsilon)\rho u^2 C_{\mathrm{D}}}{2\beta''\epsilon^3 d_{\mathrm{p}}} \tag{6.10}$$

Substituting the expression for C_{D} from Eq. (6.8) and using the definition of Re give

$$-\frac{\partial p}{\partial x} - \rho g_{\mathrm{x}} - \frac{\lambda''(1-\epsilon)\rho\mu_0 u}{\beta''\epsilon^2 d_{\mathrm{p}}^2}\left[1 - 0.74\frac{\lambda u}{d_{\mathrm{p}}}\right] = 0 \tag{6.11}$$

The quantity $\frac{\beta''\epsilon^2}{\lambda''(1-\epsilon)}$ is dependent only on the pore geometry which would be constant for each system and can be replaced by a single coefficient C_{E}. The quantity $C_{\mathrm{E}}d_{\mathrm{p}}^2$ is recognized as the intrinsic permeability and can be written as

$$K' = C_{\mathrm{E}}d_{\mathrm{p}}^2 \tag{6.12}$$

Expressions for K' in the form given in Eq. (6.12) for constant viscosity fluids are well-known, and hence an expression for C_{E} can be written as follows:

$$C_{\mathrm{E}} = \frac{1}{150}\frac{\epsilon^3}{(1-\epsilon)^2} \tag{6.13}$$

Combining Eqs. (6.11) and (6.12) gives

$$-\frac{\partial p}{\partial x} - \rho g_{\mathrm{x}} = \frac{\mu_0 u}{K'} - 0.74\frac{C_{\mathrm{E}}^{\frac{1}{2}}\mu_0\lambda u^2}{K'^{\frac{3}{2}}} \tag{6.14}$$

For $\lambda = 0$, the above equation gives the familiar Darcy equation for Newtonian fluids which is often used for solving Darcy flow problems [see, e.g., the review articles (Cheng, 1978; Combarnous & Bories, 1975; Kafoussias, 1990; Tien & Vafai, 1990)].

Darcy Natural Convection

It is assumed that the geometrical configuration is a semi-infinite vertical flat plate as shown in Fig. 6.2. The wall surface temperature T_w is considered to be constant and higher than the ambient constant temperature T_e. The flow is considered to be occurring under natural convection conditions.

Equation (6.14), when combined with the Boussinesq approximation, for the present case of natural convection in a porous medium saturated with an elastic fluid having constant viscosity, can be written as

$$u\left[1 - 0.74\frac{C_E^{\frac{1}{2}}\lambda u}{K'^{\frac{1}{2}}}\right] = \frac{K'}{\mu_0}[\rho g \beta_0 (T - T_e)] \tag{6.15}$$

The energy integral equation is given by the following:

$$\frac{d}{dx}\int_0^\delta u(T - T_e)dy = -\alpha^* \frac{\partial T}{\partial y}\bigg|_{y=0} \tag{6.16}$$

The boundary conditions on the velocity and temperature are as follows:

$$\begin{aligned} u(x,0) &= u_w; \quad u(x,\delta) = 0 \\ T(x,0) &= T_w; \quad T(x,\delta) = T_e \end{aligned} \tag{6.17}$$

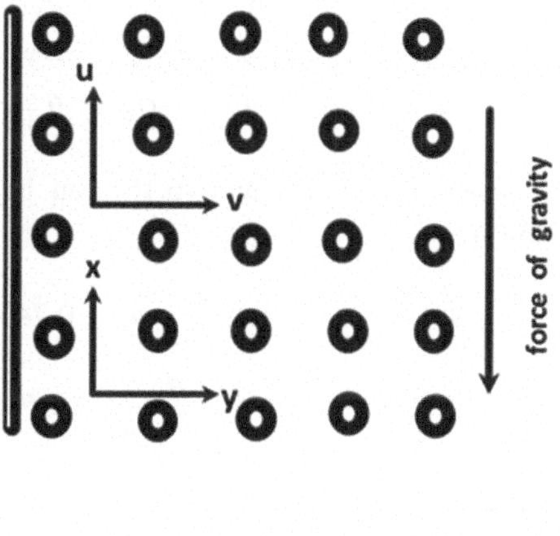

Fig. 6.2 Schematic diagram of flow past a vertical flat plate immersed in a porous medium (Shenoy, 1992). (Reprinted with permission from Elsevier and Copyright Clearance through RightsLink)

A scale analysis is now performed using $u \sim O(U_{cN})$, $x \sim O(l_c)$, $y \sim O(\delta)$, and $T - T_e \sim O(\Delta T_w = T_w - T_e)$. The energy equation based on the above scales can be written as follows:

$$\frac{U_{cN}\Delta T_w \delta}{l_c} \sim \alpha^* \frac{\Delta T_w}{\delta} \tag{6.18}$$

The characteristic velocity U_{cN} is chosen from Eq. (6.15) as follows:

$$U_{cN} \sim \frac{K'}{\mu_0}[\rho g \beta_0 \Delta T_w] \tag{6.19}$$

Solving the scales given by Eqs. (6.18) and (6.19) leads to the following

$$\frac{\delta}{l_c} \sim \frac{1}{\mathrm{Ra}_c^{\frac{1}{2}}} \tag{6.20}$$

where

$$\mathrm{Ra}_c = \frac{l_c}{\alpha^*}\left(\frac{K'\rho g \beta_0 \Delta T_w}{\mu_0}\right) \tag{6.21}$$

Since there is no characteristic length for the external flow being considered, l_c is chosen such that $\mathrm{Ra}_c = 1$ (Hellums & Churchill, 1964). Thus,

$$l_c = \alpha^* \left(\frac{\mu_0}{K'\rho g \beta_0 \Delta T_w}\right) \tag{6.22}$$

The nondimensional variables can now be written as:

$$\begin{aligned} x_1 &= x/l_c; \quad y_1 = y/l_c \\ \delta_1 &= \delta/l_c; \quad \eta = y_1/\delta_1 \\ u_1 &= u/U_{cN}; \quad \theta = (T - T_e)/(T_w - T_e) \end{aligned} \tag{6.23}$$

Thus Eqs. (6.15) and (6.16) can be written in the nondimensional forms as follows:

$$u_1 - 0.74\mathrm{We}_{D,N} u_1^2 = \theta \tag{6.24}$$

$$\frac{d}{dx_1}\int_0^1 \delta_1 u_1 \theta d\eta = -\frac{1}{\delta_1}\frac{\partial \theta}{\partial \eta}\bigg|_{\eta=0} \tag{6.25}$$

where

$$\mathrm{We}_{\mathrm{D,N}} = \left(\frac{C_{\mathrm{E}}^{\frac{1}{2}} K'^{1/2} \lambda \rho g \beta_0 \Delta T_{\mathrm{w}}}{\mu_0} \right) \tag{6.26}$$

Equations (6.24) and (6.25) are now solved subject to the following boundary conditions:

$$\begin{aligned} u_1(x_1,0) &= 1; \quad u_1(x_1,\delta_1) = 0 \\ \theta(x_1,0) &= 1; \quad \theta(x_1,\delta_1) = 0 \end{aligned} \tag{6.27}$$

Combining Eqs. (6.24) and (6.25) gives

$$\frac{d}{dx_1} \int_0^1 \delta_1 \{ u_1^2 - 0.74 \mathrm{We}_{\mathrm{D,N}} u_1^3 \} d\eta = - \frac{1}{\delta_1} \frac{\partial \theta}{\partial \eta} \bigg|_{\eta=0} \tag{6.28}$$

The velocity and temperature profiles are now to be specified. It should be noted that besides the above boundary conditions, the profiles must satisfy the following conditions for smoothness at the edges of the boundary layer:

$$\frac{\partial u_1}{\partial \eta}\bigg|_{\eta=1} = 0 \quad \frac{\partial \theta}{\partial \eta}\bigg|_{\eta=1} = 0 \tag{6.29}$$

The velocity and temperature profiles are assumed to be of the following form:

$$u_1 = \sum\nolimits_{i=0}^{i} a_i \eta^i \tag{6.30}$$

$$\theta = \sum\nolimits_{i=0}^{i} b_i \eta^i \tag{6.31}$$

It can be readily established that for satisfying the conditions given by Eqs. (6.27) and (6.29) the least that is needed is $i = 2$. The following profiles are thus chosen

$$u_1 = 1 - 2\eta + \eta^2 \tag{6.32}$$

$$\theta = 1 - 2\eta + \eta^2 \tag{6.33}$$

Substituting the above profiles in Eq. (6.28) and solving gives

$$\frac{d}{dx_1} \delta_1 \left\{ \frac{1}{5} - \frac{0.74}{7} \mathrm{We}_{\mathrm{D,N}} \right\} = \frac{2}{\delta_1} \tag{6.34}$$

For solving the above equation, it is assumed that

$$\delta_1 = B_1 x_1^{\bar{r}} \tag{6.35}$$

and similarity is seen to exist when $\bar{r} = 1/2$. Thus,

$$B_1 = \frac{2}{\{(0.2) - (0.1057)\mathrm{We}_{\mathrm{D,N}}\}^{1/2}} \tag{6.36}$$

The local Nusselt number is defined as

$$\mathrm{Nu}_{\mathrm{x,D,N}} = -\frac{x_1}{\delta_1}\left(\frac{\partial \theta}{\partial \eta}\right)\bigg|_{\eta=0} \tag{6.37}$$

$$= \frac{2}{B_1} x_1^{1/2} \tag{6.38}$$

Thus

$$\mathrm{Nu}_{\mathrm{x,D,N}} = 0.4472\{1 - (0.5285)\mathrm{We}_{\mathrm{D,N}}\}^{1/2}\mathrm{Ra}_{\mathrm{x,D,N}}^{1/2} \tag{6.39}$$

where

$$\mathrm{Ra}_{\mathrm{x,D,N}} = \frac{x}{\alpha^*}\left(\frac{K'\rho g \beta_0 \Delta T_{\mathrm{w}}}{\mu_0}\right) \tag{6.40}$$

The first step is to check the propriety of Eq. (6.39) for Newtonian fluids, i.e., when $\mathrm{We}_{\mathrm{D,N}} = 0$. The predictions for Newtonian fluids when compared with those available in literature as shown in Table 6.1 give the closest match to the exact solution values, thus instilling confidence in the solution procedure employed in the above analysis.

It should be noted that Eq. (6.39) is valid for all values of $\mathrm{We}_{\mathrm{D,N}} \leq 0.7$. For $0.7 \leq \mathrm{We}_{\mathrm{D,N}} \leq 2.0$, the asymptotic value of the expression is determined by putting $\mathrm{We}_{\mathrm{D,N}} = 0.7$. The effect of the elasticity is brought out by a plot in Fig. 6.3.

Table 6.1 Comparison of the local Nusselt number for Darcy natural convection flow predictions of the approximate integral solutions with other exact and approximate solutions for Newtonian fluids in the case of the vertical flat plate at constant temperature (Shenoy, 1992)

	$\mathrm{Nu}_{\mathrm{x,D,N}}/(\mathrm{Ra}_{\mathrm{x,D,N}})^{1/2}$ (Newtonian fluids)			
$\mathrm{We}_{\mathrm{D,N}}$	(Shenoy, 1992) Approximate integral solution	(Chen & Chen, 1988a) Exact solution	(Wang & Tu, 1989) Exact solution	(Nakayama & Koyama, 1991) Approximate solution
0	0.4472	0.4437	0.4440	0.4082

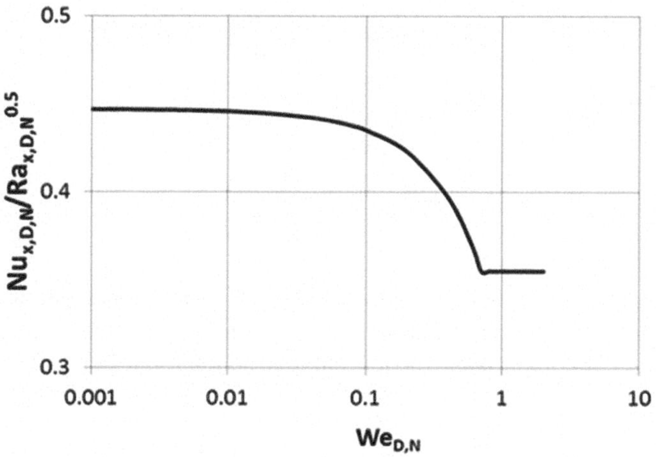

Fig. 6.3 Variation of the local Nusselt number with elasticity for Darcy natural convection flow (Shenoy, 1992). (Reprinted with permission from Elsevier and Copyright Clearance through RightsLink)

It can be seen that for values of $We_{D,N}$ up to 0.1, the heat transfer rates deviate from the Newtonian value by less than 5%; however, at $We_{D,N}$ beyond 0.1, there is a rapid drop in the heat transfer rate due to the dominant elastic effects up to $We_{D,N}$ of 0.7, after which it attains a constant value.

Darcy Forced Convection

It is assumed that the flow takes place by forced convection flow with a uniform parallel velocity of u_e past an isothermal semi-infinite vertical flat plate as shown in Fig. 6.2. Equation (6.14) gives the momentum equation which is rewritten as:

$$u_e\left[1 - 0.74\frac{C_E^{\frac{1}{2}}\lambda u_e}{K'^{\frac{1}{2}}}\right] = \frac{K'}{\mu_0}\left(-\frac{\partial p}{\partial x} - \rho g\right) \tag{6.41}$$

The energy integral equation for the flow under consideration is given by the following:

$$\frac{d}{dx}\int_0^{\delta} u_e(T - T_e)dy = -\alpha^* \frac{\partial T}{\partial y}\bigg|_{y=0} \tag{6.42}$$

The characteristic velocity U_{cF} for this case is defined as follows:

$$U_{cF} = u_e[1 - 0.74\mathrm{We}_{D,F}] = \frac{K'}{\mu_0}\left(-\frac{\partial p}{\partial x} - \rho g\right) \tag{6.43}$$

where

$$\mathrm{We}_{D,F} = \left(\frac{C_E^{\frac{1}{2}}\mu_0 u_e}{K'^{1/2}}\right) \tag{6.44}$$

Since Eqs. (6.41) and (6.42) are decoupled unlike in the earlier natural convection case, the solution of the above equations is rather simple. Hence, details are not provided here. The expression for the Nusselt number can be easily obtained as given below.

$$\mathrm{Nu}_{x,D,F} = 0.5774\{1 - (0.74)\mathrm{We}_{D,F}\}^{1/2}\mathrm{Pe}_{x,D,F}^{1/2} \tag{6.45}$$

where

$$\mathrm{Pe}_{x,D,F} = \frac{xu_e}{\alpha^*} \tag{6.46}$$

When $\mathrm{We}_{D,F} = 0$ in Eq. (6.45), the coefficient obtained is 0.5774 for Newtonian fluids. For the isothermal flat vertical plate, it has been shown (Bejan, 1984; Nakayama & Pop, 1991) that this coefficient is equal to 0.5641 if an exact similarity solution is obtained. The coefficient obtained in the above analysis differs by only about 2.3%. It should be noted that Eq. (6.45) is valid for all values of $\mathrm{We}_{D,F} \leq 0.7$. For $0.7 \leq \mathrm{We}_{D,F} \leq 2.0$, the asymptotic value of the expression is determined by putting $\mathrm{We}_{D,F} = 0.7$. Figure 6.4 shows the effect of elasticity on the heat transfer characteristics during forced convection. It can be seen that the heat transfer reduction due to elasticity is much more pronounced for forced convection than for natural convection.

Darcy Mixed Convection

It is well-known that in any heat transfer situation, density differences are bound to arise, and a forced field is likely to be superimposed by natural convection effects. In forced convection, if the momentum transport rates are significantly strong, then the effects of natural convection can be neglected. Similarly, if the buoyancy forces are of relatively greater magnitude, then the effects of forced convection flow may be ignored. But in many practical heat transfer situations, the forced and the natural convection effects are of comparable order, and hence it is the combined effect of this mixed convection that truly determines the heat transfer rate. Mixed convection

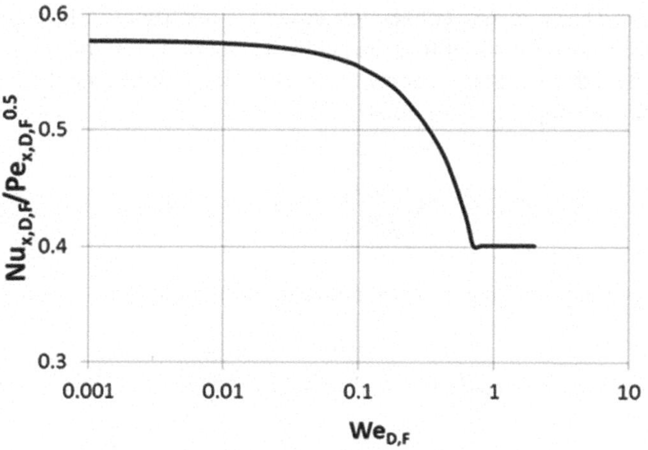

Fig. 6.4 Variation of the local Nusselt number with elasticity for Darcy forced convection flow (Shenoy, 1992). (Reprinted with permission from Elsevier and Copyright Clearance Center through RightsLink)

problems are also as difficult to solve as the natural convection, again due to the coupling of the momentum and energy equations.

It is assumed that the geometrical configuration is a vertical flat plate with a constant wall surface temperature T_w which is higher than the ambient constant temperature at the edge of the boundary-layer T_e. The flow is occurring under the combined effect of forced and natural convection conditions. The external velocity u_e for the forced convection is considered to be uniform and parallel. Equation for the present case of mixed convection in porous medium saturated with an elastic fluid of constant viscosity can be written as

$$u\left[1 - 0.74\frac{C_E^{\frac{1}{2}}\lambda u}{K'^{\frac{1}{2}}}\right] = \frac{K'}{\mu_0}\left(-\frac{\partial p}{\partial x} - \rho g\right) + \frac{K'}{\mu_0}[\rho g \beta_0 (T - T_e)] \tag{6.47}$$

At the edge of the boundary layer, $u = u_e$ and $T = T_e$. Thus, the above Eq. (6.47) can be written as

$$u_e\left[1 - 0.74\frac{C_E^{\frac{1}{2}}\lambda u_e}{K'^{\frac{1}{2}}}\right] = \frac{K'}{\mu_0}\left(-\frac{\partial p}{\partial x} - \rho g\right) \tag{6.48}$$

Equation (6.47) is rewritten using Eq. (6.48) as

$$u\left[1 - 0.74\frac{C_E^{\frac{1}{2}}\lambda u}{K'^{\frac{1}{2}}}\right] = u_e\left[1 - 0.74\frac{C_E^{\frac{1}{2}}\lambda u_e}{K'^{\frac{1}{2}}}\right] + \frac{K'}{\mu_0}[\rho g \beta_0 (T - T_e)] \tag{6.49}$$

Solving the Darcy mixed convection case is a nontrivial task. Hence an extra assumption is needed for simplification before a solution is sought. It is assumed that the velocity field due to natural convection is not altered much and the left-hand side of Eq. (6.49) can simply be written as:

$$u\left[1-0.74\frac{C_{\mathrm{E}}^{\frac{1}{2}}\lambda u}{K'^{\frac{1}{2}}}\right]=u[1-0.74\mathrm{We}_{\mathrm{D,N}}] \tag{6.50}$$

The energy integral equation is given by the following:

$$\frac{d}{dx}\int_{0}^{\delta}u(T-T_{\mathrm{e}})dy=-\alpha^{*}\frac{\partial T}{\partial y}\bigg|_{y=0} \tag{6.51}$$

The boundary conditions on the velocity and temperature are as follows:

$$\begin{aligned} u(x,0)&=u_{\mathrm{w}}; \quad u(x,\delta)=u_{\mathrm{e}} \\ T(x,0)&=T_{\mathrm{w}}; \quad T(x,\delta)=T_{\mathrm{e}} \end{aligned} \tag{6.52}$$

A scale analysis is now performed using $u\sim O(U_{\mathrm{cM}})$, $x\sim O(l_{\mathrm{c}})$, $y\sim O(\delta)$, and $T\sim T_{\mathrm{e}}\sim O(\Delta T\mathrm{w}=T_{\mathrm{w}}-T_{\mathrm{e}})$. The energy equation based on the above scales can be written as follows:

$$\frac{U_{\mathrm{cM}}\Delta T_{\mathrm{w}}\delta}{l_{\mathrm{c}}}\sim\alpha^{*}\frac{\Delta T_{\mathrm{w}}}{\delta} \tag{6.53}$$

From Eq. (6.50), U_{cM} is chosen as follows:

$$U_{\mathrm{cM}}[1-0.74\mathrm{We}_{\mathrm{D,N}}]\sim u_{\mathrm{e}}[1-0.74\mathrm{We}_{\mathrm{D,F}}]+\frac{K'}{\mu_{0}}[\rho g\beta_{0}(T-T_{\mathrm{e}})] \tag{6.54}$$

Solving the scales given by Eq. (6.53) and (6.54) leads to the following

$$\frac{\delta}{l_{\mathrm{c}}}\sim\frac{1}{\mathrm{Pe}_{\mathrm{c,M}}^{1/2}} \tag{6.55}$$

where

$$\mathrm{Pe}_{\mathrm{c,M}}=\frac{l_{\mathrm{c}}}{\alpha^{*}}U_{\mathrm{cM}} \tag{6.56}$$

$$U_{cM} \sim \frac{\alpha^*}{x} Pe_{x,D,F} \frac{[1 - 0.74We_{D,F}]}{[1 - 0.74We_{D,N}]} \left\{ 1 + \frac{Ra_{x,D,N}}{Pe_{x,D,F}[1 - 0.74We_{D,F}]} \right\} \tag{6.57}$$

Since there is no characteristic length for the external flow being considered, l_c is chosen such that $Pe_{cM} = 1$ (Hellums & Churchill, 1964). Thus,

$$l_c = \frac{\alpha^*}{U_{cM}} \tag{6.58}$$

The nondimensional variables can now be written as:

$$\begin{aligned} x_1 &= x/l_c; \quad y_1 = y/l_c \\ \delta_1 &= \delta/l_c; \quad \eta = y_1/\delta_1 \\ u_1 &= u/U_{cM}; \quad \theta = (T - T_e)/(T_w - T_e) \end{aligned} \tag{6.59}$$

Thus, Eqs. (6.50) and (6.51) can be written in the nondimensional forms as follows:

$$u_1 = \frac{\{1 + (Ra_{x,D,N}/Pe_{x,D,F}[1 - 0.74We_{D,F}])\theta\}}{\{1 + (Ra_{x,D,N}/Pe_{x,D,F}[1 - 0.74We_{D,F}])\}} \tag{6.60}$$

$$\frac{d}{dx_1} \int_0^1 \delta_1 u_1 \theta d\eta = -\frac{1}{\delta_1} \frac{\partial \theta}{\partial \eta}\bigg|_{\eta=0} \tag{6.61}$$

Combining Eqs. (6.60) and (6.61) gives

$$\frac{d}{dx_1} \int_0^1 \delta_1 \frac{\{1 + (Ra_{x,D,N}/Pe_{x,D,F}[1 - 0.74We_{D,F}])\theta\}}{\{1 + (Ra_{x,D,N}/Pe_{x,D,F}[1 - 0.74We_{D,F}])\}} \theta d\eta = -\frac{1}{\delta_1} \frac{\partial \theta}{\partial \eta}\bigg|_{\eta=0} \tag{6.62}$$

The temperature profile used earlier for the pure forced as well as pure natural convection cases and given by Eq. (6.33) can be used in the case of mixed convection as well. Integration of Eq. (6.62) is straightforward, and the procedure identical to that used for the natural convection case is repeated, and an expression for Nusselt number is obtained as follows:

$$\begin{aligned} Nu_{x,D,M} = \{&[1/3][1 - (0.74)We_{D,F}]Pe_{x,D,F} \\ &+ [1/5][1 - (0.74)We_{D,N}]Ra_{x,D,N}\}^{1/2} \end{aligned} \tag{6.63}$$

or

$$Nu^2_{x,D,M} = Nu^2_{x,D,F} + Nu^2_{x,D,N} \tag{6.64}$$

The entire second term on the right-hand side of Eq. (6.63) would have been identical to the expression of Nusselt number in Eq. (6.39) for the pure natural convection case but for the coefficient 0.74. In the above Eq. (6.63), the term $[1\text{–}0.74We_{D,N}]$ appears because of the simplifying assumption used in Eq. (6.50) and hence needs to be corrected to $(1\text{–}0.5285We_{D,N})$ in order that Eq. (6.64) holds exactly.

It is worth noting that Eq. (6.64) is of the form similar to those of the correlating equations for combined laminar forced and free convection heat transfer (Churchill, 1977; Ruckenstein, 1978) for Newtonian fluids and (Shenoy, 1980a, 1980b) for non-Newtonian fluids in homogeneous media. Such equations which interpolate the two extremes of forced and free convection have been shown to give reasonably accurate results. In the case of convective flow through porous media, too, such correlating equations are very effective (Shenoy, 1992, 1993).

The predictions of Eq. (6.64) for the Newtonian case, i.e., $We_{D,N} = We_{D,F} = 0$, are compared with those available in the literature in Table 6.2 and found to give very close agreement to the results from exact solution.

For values of $Ra_{x,D,N}/Pe_{x,D,F}$ greater than zero, it can be seen that the heat transfer rates increase. However, with increasing elasticity, the heat transfer rates are reduced as can be seen from Fig. 6.5. The asymptotic value of the expression is obtained by putting $We_{D,N} = We_{D,F} = 0.7$.

Table 6.2 Comparison of approximate and exact predictions of local Nusselt number for Darcy mixed convection flow of Newtonian fluids on vertical flat plate at constant temperature (Shenoy, 1992)

		$Nu_{x,D,M}/(Pe_{x,D,F})^{1/2}$ (Newtonian fluids)	
$We_{D,N}$ and $We_{D,F}$	$Ra_{x,D,N}/Pe_{x,D,F}$	(Shenoy, 1992) Approximate integral	(Nakayama & Shenoy, 1993) Exact
0	0	0.5774	0.5641
	0.5	0.6583	0.6473
	1.0	0.7303	0.7205
	15	1.826	1.812
	50	3.215	3.140
	100	4.509	4.443
	200	6.351	6.279
	400	8.963	8.886

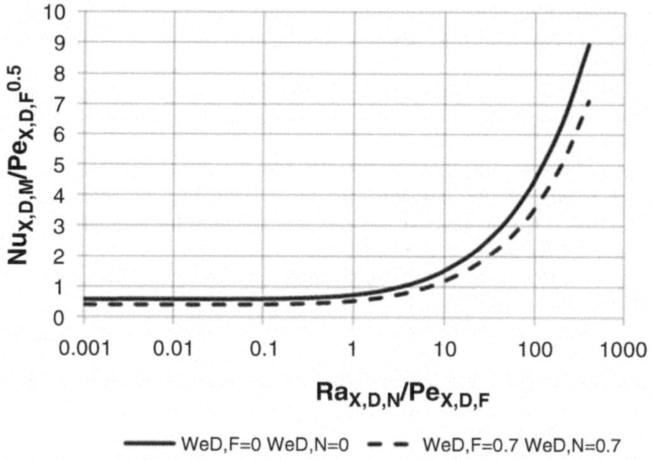

Fig. 6.5 Variation of the local Nusselt number with elasticity for Darcy mixed convection flow (Shenoy, 1992). (Reprinted with permission from Elsevier and Copyright Clearance Center through RightsLink)

References

Bejan, A. (1984). *Convective heat transfer*. New York: Wiley.

Bejan, A. (1987). In S. Kakac, R. K. Shah, & W. Aung (Eds.), *Handbook of single-phase convective heat transfer, convective heat transfer in porous media*. New York: Wiley.

Bejan, A., & Poulikakos, D. (1984). The non-Darcy regime for vertical boundary layer natural convection in a porous medium. *International Journal of Heat and Mass Transfer, 27*(5), 717–722.

Boger, D. V. (1977/78). A highly elastic constant viscosity fluid. *Journal of Non-Newtonian Fluid Mechanics, 3*, 87–91.

Brinkman, H. C. (1947). A calculation of the viscous force exerted by a flowing fluid on a dense swarm of particles. *Applied Scientific Research, A1*, 27–34.

Broadbent, J. M., & Mena, B. (1974). Slow flow of an elastico-viscous fluid past cylinders and spheres. *Chemical Engineering Journal, 8*(1), 11–19.

Cao, Y., & Cui, X. (2015). Natural convection of power-law fluids in porous media with variable thermal and mass diffusivity. *International Journal of Heat and Technology, 33*(2), 85–90.

Caswell, B., & Schwarz, W. H. (1962). The creeping motion of a non-newtonian fluid past a sphere. *Journal of Fluid Mechanics, 13*(3), 417–426.

Chamkha, A. J., & Ben-Nakhi, A. (2007). Coupled heat and mass transfer in mixed convection flow of a non-newtonian fluidover a permeable surface embedded in a non-darcian porous medium. *International Journal of Heat and Technology, 25*(1), 33–41.

Chen, H.-T., & Chen, C.-K. (1988a). Free convection of non-newtonian fluids along a vertical surface embedded in a porous medium. *Transactions ASME, Journal Heat Transfer, 110*, 257–260.

Chen, H.-T., & Chen, C.-K. (1988b). Natural convection of a non-newtonian fluid about a horizontal cylinder and a sphere in a porous medium. *International Communications in Heat and Mass Transfer, 15*, 605–614.

Cheng, P. (1978). Heat transfer in geothermal systems. *Advances in Heat Transfer, 14*, 1–105.

Chhabra, R. P. (1986). Encyclopedia of fluid mechanics, vol. 1, ch. 30. In N. P. Cheremisinoff (Ed.), *Steady non-newtonian flow about a rigid sphere* (pp. 983–1033). Houston, TX: Gulf Publishing Co.

Chhabra, R. P., Uhlherr, P. H., & Boger, D. V. (1980). The influence of fluid elasticity on the drag coefficient for creeping flow around a sphere. *Journal of Non-Newtonian Fluid Mechanics, 6* (3–4), 187–199.

Choplin, L., Carreau, P., & Aitkadi, A. (1983). Highly elastic constant viscosity fluids. *Polymer Engineering & Science, 23*(8), 459–464.

Churchill, S. W. (1977). A comprehensive correlating equation for laminar, assisting, forced and free convection. *AICHE Journal, 23*(1), 10–16.

Combarnous, M. A., & Bories, S. A. (1975). Hydrothermal convection in saturated porous media. *Advances Hydroscience, 10*, 231–307.

Ergun, S. (1952). Fluid flow through packed columns. *Chemical Engineering Progress, 48*(2), 89–94.

Fand, R. M., Steinberger, T. E., & Cheng, P. (1986). Combined natural and forced convection heat transfer from a horizontal cylinder embedded in a porous medium. *International Journal of Heat and Mass Transfer, 29*(1), 119–133.

Forchheimer, P. (1901). Wasserbewegung durch boden. *Zeitschrift des Vereins Deutscher Ingenieure, 45*, 1782–1788.

Giesekus, H. (1963). Die simultane translations- und rotationsbewegung einer kugel in einer elastikoviskosen flussigkeit. *Rheologica Acta, 3*, 59–71.

Hellums, J. D., & Churchill, S. W. (1964). Simplification of the mathematical description of boundary and initial value problems. *AICHE Journal, 10*(1), 110–114.

Ingham, D. B. (1986). The non-darcy free convection boundary layer on axisymmetric and two-dimensional bodies of arbitrary shape. *International Journal of Heat and Mass Transfer, 29*, 1759–1763.

Kafoussias, N. G. (1990). Principles of flow through porous media with heat transfer. In *Encyclopedia of fluid mechanics, vol. 10, ch. 20* (pp. 663–686). London: Gulf Publishing Co.

Lauriat, G., & Prasad, V. (1987). Natural convection in a vertical porous cavity: A numerical study for brinkman-extended darcy formulation. *Transactions ASME Journal of Heat Transfer, 109* (3), 688–696.

Leslie, F. M., & Tanner, R. I. (1961). The slow flow of a visco-elastic liquid past a sphere. *The Quarterly Journal of Mechanics and Applied Mathematics, 14*(1), 36–48.

Mena, B., & Caswell, B. (1974). Slow flow of an elastic-viscous fluid past an immersed body. *Chemical Engineering Journal, 8*(2), 125–134.

Metzner, A. B., White, J. L., & Denn, M. M. (1966). Constitutive equations for viscoelastic fluids for short deformation periods and for rapidly changing flows: Significance of the Deborah number. *AICHE Journal, 12*(5), 863–866.

Muskat, M. (1946). *The flow of homogeneous fluids through porous media*. Ann Arbor, MI: Edwards.

Nakayama, A., & Koyama, H. (1991). Buoyancy-induced flow of non-newtonian fluids over a non-isothermal body of arbitrary shape in a fluid-saturated porous medium. *Applied Scientific Research, 48*(1), 55–70.

Nakayama, A., & Pop, I. (1991). A unified similarity transformation for free, forced and mixed convection in darcy and non-darcy porous media. *International Journal of Heat and Mass Transfer, 34*, 357–367.

Nakayama, A., & Shenoy, A. V. (1992). A unified similarity transformation for darcy and non-darcy forced, free and mixed convection heat transfer in non-newtonian inelastic fluid-saturated porous media. *Chemical Engineering Journal, 50*(1), 33–45.

Nakayama, A., & Shenoy, A. V. (1993). Combined forced and free convection heat transfer in non-newtonian fluid saturated porous medium. *Applied Scientific Research, 50*(1), 83–95.

Nield, D. A., & Joseph, D. D. (1985). Effects of quadratic drag on convection in a saturated porous medium. *Physics of Fluids, 28*(3), 995–997.

Pascal, H. (1990a). Non-isothermal flow of non-newtonian fluids through a porous medium. *International Journal of Heat and Mass Transfer, 33*(9), 1937–1944.

Pascal, H. (1990b). Some self-similar two-phase flows of non-newtonian fluids through a porous medium. *The journal Studies in Applied Mathematics, 82*, 305–318.

Pascal, H., & Pascal, J. P. (1989). Non-linear effects of non-newtonian fluids on natural convection in a porous medium. *Physica D, 40*(3), 393–402.

Plumb, O. A., & Huenefeld, J. C. (1981). Non-darcy natural convection from heated surfaces in saturated porous medium. *International Journal of Heat and Mass Transfer, 24*(4), 765–768.

Poulikakos, D., & Bejan, A. (1985). The departure from darcy flow in natural convection in a vertical porous layer. *Physics of Fluids, 28*(12), 3477–3484.

Ruckenstein, E. (1978). Interpolating equations between two limiting cases for the heat transfer. *AICHE Journal, 24*(5), 940–941.

Rumer, R. R. (1969). In R. J. De Wiest (Ed.), *Flow through porous media*. Academic Press: New York.

Shenoy, A. V. (1980a). A correlating equation for combined laminar forced and free convection heat transfer to power-law fluids. *AICHE Journal, 26*(3), 505–507.

Shenoy, A. V. (1980b). Combined laminar forced and free convection heat transfer to viscoelastic fluids. *AICHE Journal, 26*(4), 683–685.

Shenoy, A. V. (1986). Handbook of heat mass transfer, vol. 1, ch. 5. In N. P. Cheremisinoff (Ed.), *Natural convection heat transfer to power-law fluids* (pp. 183–210). Houston, TX: Gulf Publishing Co.

Shenoy, A. V. (1988). Encyclopedia of fluid mechanics, vol. 7, ch. 10. In N. P. Cheremisinoff (Ed.), *Natural convection heat transfer to viscoelastic fluids* (pp. 287–304). Houston, TX: Gulf Publishing Co.

Shenoy, A. V. (1992). Darcy natural, forced and mixed convection heat transfer from an isothermal vertical flat plate embedded in a porous medium saturated with an elastic fluid of constant viscosity. *International Journal of Engineering Science, 30*(4), 455–467.

Shenoy, A. V. (1993). Darcy-forchheimer natural, forced and mixed convection heat transfer in non-newtianian power-law fluid-saturated porous media. *Transport in Porous Media, 11*(3), 219–241.

Shenoy, A. V. (1994). Non-newtonian fluid heat transfer in porous media. *Advanced Heat Transfer, 24*, 101–190.

Tien, C. L., & Vafai, K. (1990). Convective and radiative heat transfer in porous media. *Advances in Applied Mechanics, 27*, 225–281.

Tong, T. W., & Subramanian, E. (1985). A boundary layer analysis for natural convection in vertical porous enclosures – use of the Brinkman-extended Darcy model. *International Journal of Heat and Mass Transfer, 28*(3), 563–571.

Trevisan, O. V., & Bejan, A. (1990). Combined heat and mass transfer by natural convection in a porous medium. *Advances Heat Transfer, 20*, 315–352.

Vasantha, R., Pop, I., & Nath, G. (1986). Non-darcy natural convection over a slender vertical frustum of a cone in a saturated porous medium. *International Journal of Heat and Mass Transfer, 29*(1), 153–156.

Wang, C., & Tu, C. (1989). Boundary layer flow and heat transfer of non-newtonian fluids in porous media. *International Journal of Heat and Fluid Flow, 10*(2), 160–165.

Wang, C., Tu, C., & Zhang, X. (1990). Mixed convection of non-newtonian fluids from a vertical plate embedded in a porous medium. *Acta Mechanica Sinica, 6*(3), 214–220.

Ward, J. C. (1964). Turbulent flow in porous media. *Journal of the Hydraulics Division ASCE, 90* (5), 1–12.

White, J. L. (1964). Dynamics of viscoelastic fluids, melt fracture, and the rheology of fiber spinning. *Journal of Applied Polymer Science, 8*(5), 2339–2357.

Chapter 7
Forced Convection Heat Transfer in Internal Flows Through Porous Media

Channel Flow

Heat transfer to non-Newtonian fluids in saturated porous medium for the well-known Darcy flow (Cao & Cui, 2015; Chen & Chen, 1988a, 1988b; Nakayama & Koyama, 1991; Nakayama & Shenoy, 1993a; Pascal, 1990a, 1990b; Pascal & Pascal, 1989; Wang & Tu, 1989; Wang, Tu, & Zhang, 1990), the various cases of Darcy-Forchheimer flow (Nakayama & Shenoy, 1992; Shenoy, 1993a), as well as the Darcy-Brinkman flow (Nakayama & Shenoy, 1993b) all have treated only non-Newtonian inelastic fluids. In non-Newtonian fluid mechanics, the effects of elasticity are of considerable interest, and this has been studied for the simplest external flow past a vertical flat plate (Shenoy, 1993a) and internal flow through a channel (Shenoy, 1993b) embedded in a porous medium saturated with an elastic fluid. The focus of this section is on the internal channel flow.

The flow is assumed to be confined within parallel walls subjected to uniform heat flux and immersed in a porous medium saturated with an elastic fluid of constant viscosity (Boger, 1977/78; Choplin, Carreau, & Aitkadi, 1983). This helps in isolating the effect of elasticity. If a viscoelastic fluid showing shear-thinning characteristics is treated, then the effects of elasticity and shear-thinning often overlap, and individual effects cannot be deciphered.

Appropriate transformation of variables is done along the lines for the Newtonian case (Nakayama, Koyama, & Kuwahara, 1988), and a solution based on the approximate integral method is obtained.

It is assumed that the geometrical configuration is a two-dimensional channel embedded in a medium of relatively high porosity shown in Fig. 7.1.

The solid matrix and the fluid are assumed to be in thermal equilibrium so that they could be treated as a continuum. It is assumed that the fluid is elastic in nature but shows a constant viscosity behavior (Boger, 1977/78; Choplin et al., 1983). For such a fluid, the modification required in the Darcy model has been presented (Shenoy, 1993b).

A. Shenoy, *Rheology of Drag Reducing Fluids*,
https://doi.org/10.1007/978-3-030-40045-3_7

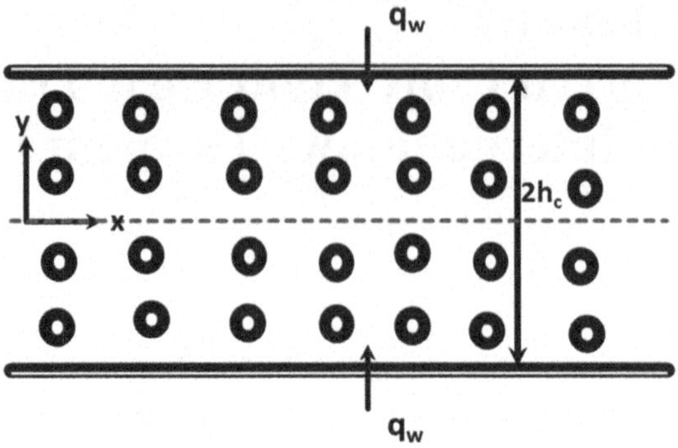

Fig. 7.1 Physical model and its coordinates for internal channel flow (Shenoy, 1993b). (Reprinted with permission from Springer and Copyright Clearance through RightsLink)

The governing equations, namely, the Brinkman-extended Darcy model and the energy equation, can thus be written as follows:

$$\frac{\mu_0}{\epsilon}\frac{d^2u}{dy^2} - \frac{\mu_0 u}{K'} + 0.74\frac{C_{\mathrm{E}}^{\frac{1}{2}}\mu_0\lambda u^2}{K'^{\frac{3}{2}}} = \frac{\partial p}{\partial x} \tag{7.1}$$

$$\rho C_{\mathrm{p}} u\frac{\partial T}{\partial x} = k\frac{\partial^2 T}{\partial y^2} \tag{7.2}$$

Since the velocity and temperature fields are symmetric about the channel center line, only the upper half of the channel is taken into consideration.

u is the Darcian (apparent) velocity in the x-direction; p and T are the local pressure and temperature; ρ, C_{p}, and μ_0 are the density, specific heat, and viscosity of the elastic fluid; ϵ is the porosity; K' is the permeability; and k is the effective thermal conductivity of the fluid-saturated porous medium.

The above governing equations are subject to the following boundary conditions:

$$\begin{aligned} &\text{For } y = 0: \frac{du}{dy} = 0; \quad \frac{\partial T}{\partial y} = 0 \\ &\text{For } y = h_{\mathrm{c}}: u = 0; \quad \frac{\partial T}{\partial y} = \frac{q_{\mathrm{w}}}{k} \end{aligned} \tag{7.3}$$

It is assumed that the upper and lower impermeable walls are subjected to constant heat flux q_{w}.

Equation (7.1) can be written in nondimensional form as follows:

$$\frac{d^2u_1}{dy_1^2} - \sigma^2 u_1 + 0.74\mathrm{We}\sigma^2 u_1^2 = \frac{\epsilon h^2}{\mu_0 u_{\mathrm{c}}}\frac{\partial p}{\partial x} \tag{7.4}$$

where

$$u_1 = u/u_c; \quad y_1 = y/h_c$$
$$\sigma = h_c / \left(\frac{K'}{\epsilon}\right)^{\frac{1}{2}}; \quad \text{We} = \frac{C_E^{\frac{1}{2}} \lambda u_c}{K'^{\frac{1}{2}}} \tag{7.5}$$

u_c is the velocity along the duct centerline at $y = 0$. Equation (7.4) is integrated using the boundary condition in Eq. (7.3). Thus,

$$\left.\frac{du_1}{dy_1}\right|_{y_1=1} - \sigma^2 \int_0^1 u_1 dy_1 + 0.74\text{We}\sigma^2 \int_0^1 u_1^2 dy_1 = \frac{\epsilon h_c^2}{\mu_0 u_c} \frac{\partial p}{\partial x} \tag{7.6}$$

A further auxiliary relationship is obtained by writing Eq. (7.4) at the wall. Thus,

$$\left.\frac{d^2 u_1}{dy_1^2}\right|_{y_1=1} = \frac{\epsilon h^2}{\mu_0 u_c} \frac{\partial p}{\partial x} \tag{7.7}$$

where the boundary condition given by Eq. (7.3) has been used. Combining Eqs. (7.6) and (7.7) gives

$$\left.\frac{du_1}{dy_1}\right|_{y_1=1} - \sigma^2 \int_0^1 u_1 dy_1 + 0.74\text{We}\sigma^2 \int_0^1 u_1^2 dy_1 = \left.\frac{d^2 u_1}{dy_1^2}\right|_{y_1=1} \tag{7.8}$$

In line with the general tradition of an integral solution, a velocity profile is now chosen as

$$u_1 = 1 - y_1^{\zeta} \tag{7.9}$$

Using the above expression, Eq. (7.8) is solved to give the following form for the shape factor ζ:

$$\zeta = \left(\frac{1}{2}\right)\left\{1 + \left(9 + 4\sigma^2\left[1 - 0.74\text{We}\left(\frac{2\zeta}{2\zeta + 1}\right)\right]\right)^{0.5}\right\} \tag{7.10}$$

The above expression is implicit in ζ but can easily be solved by iteration using $\zeta = 2$ as a first approximation. Equation (7.10) reduces exactly to the equation for the Newtonian case (Nakayama et al., 1988) when We $= 0$.

The energy equation given by Eq. (7.2) can be integrated over the entire range $0 \leq y \leq h_c$.

Thus,

$$\rho C_p u_B h_c \frac{\partial T_B}{\partial x} = q_w \tag{7.11}$$

where u_B is the bulk mean velocity. Combining Eqs. (7.2) and (7.11) gives

$$\frac{u}{u_B} = k\frac{d^2\theta}{dy_1^2} \tag{7.12}$$

where

$$\theta = \frac{k(T - T_w)}{q_w h} \tag{7.13}$$

Upon noting that the

$$\frac{u_c}{u_B} = \frac{1}{\int_0^1 u_1 dy_1} = \frac{1+\zeta}{\zeta} \tag{7.14}$$

Equation (7.12) is rewritten using Eq. (7.9) as

$$\frac{1+\zeta}{\zeta}\left(1 - y_1^{\zeta}\right) = \frac{d^2\theta}{dy_1^2} \tag{7.15}$$

The foregoing equation is integrated twice to yield

$$\theta = \frac{1}{\zeta(\zeta+2)}\left(1 - y_1^{\zeta+2}\right) - \frac{1+\zeta}{2\zeta}\left(1 - y_1^2\right) \tag{7.16}$$

The Nusselt number can easily be obtained by integrating the product $u_1\theta$ over the channel, namely,

$$\text{Nu} = \frac{4\zeta}{(\zeta+1)}\int_0^1 u_1\theta dy_1 = \frac{12(\zeta+3)(2\zeta+3)}{2\zeta^2 + 13\zeta + 17} \tag{7.17}$$

It is worth noting that the expression for the Nusselt number for the elastic fluid considered herein is the same as that for Newtonian fluids. However, the expression for ζ differs, as can be seen from Eq. (7.10).

The Nusselt number for varying σ is plotted in Fig. 7.2.

It is seen that as $\sigma \to \infty$, the Nusselt number rightly approaches the value of 12 for all We. However, at lower values of σ, the heat transfer rate decreases with increasing elasticity and approaches Poiseuille flow values.

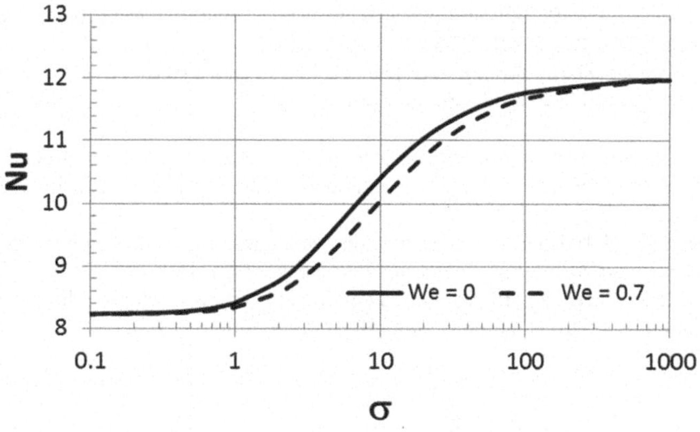

Fig. 7.2 Variation of Nusselt number with σ for different values of We (Shenoy, 1993b). (Reprinted with permission from Springer and Copyright Clearance Center through RightsLink)

References

Boger, D. V. (1977/78). A highly elastic constant viscosity fluid. *Journal of Non-Newtonian Fluid Mechanics, 3*, 87–91.

Cao, Y., & Cui, X. (2015). Natural convection of power-law fluids in porous media with variable thermal and mass diffusivity. *International Journal of Heat and Technology, 33*(2), 85–90.

Chen, H.-T., & Chen, C.-K. (1988a). Free convection of non-newtonian fluids along a vertical surface embedded in a porous medium. *Transactions ASME, Journal Heat Transfer, 110*, 257–260.

Chen, H.-T., & Chen, C.-K. (1988b). Natural convection of a non-newtonian fluid about a horizontal cylinder and a sphere in a porous medium. *International Communications in Heat and Mass Transfer, 15*, 605–614.

Choplin, L., Carreau, P., & Aitkadi, A. (1983). Highly elastic constant viscosity fluids. *Polymer Engineering & Science, 23*(8), 459–464.

Nakayama, A., & Koyama, H. (1991). Buoyancy-induced flow of non-newtonian fluids over a non-isothermal body of arbitrary shape in a fluid-saturated porous medium. *Applied Scientific Research, 48*(1), 55–70.

Nakayama, A., & Shenoy, A. V. (1992). A unified similarity transformation for darcy and non-darcy forced, free and mixed convection heat transfer in non-newtonian inelastic fluid-saturated porous media. *Chemical Engineering Journal, 50*(1), 33–45.

Nakayama, A., & Shenoy, A. V. (1993a). Combined forced and free convection heat transfer in non-newtonian fluid saturated porous medium. *Applied Scientific Research, 50*(1), 83–95.

Nakayama, A., & Shenoy, A. V. (1993b). Non-darcy forced convection heat transfer in a channel embedded in a non-newtonian inelastic fluid saturated porous medium. *Canadian Journal of Chemical Engineering, 71*(1), 168–173.

Nakayama, A., Koyama, H., & Kuwahara, F. (1988). An analysis on forced convection in a channel filled with a Brinkman-darcy porous medium: Exact and approximate solutions. *Warme- und Stoffubertragung, 23*, 291–295.

Pascal, H. (1990a). Non-isothermal flow of non-newtonian fluids through a porous medium. *International Journal of Heat and Mass Transfer, 33*(9), 1937–1944.

Pascal, H. (1990b). Some self-similar two-phase flows of non-newtonian fluids through a porous medium. *The journal Studies in Applied Mathematics, 82*, 305–318.

Pascal, H., & Pascal, J. P. (1989). Non-linear effects of non-newtonian fluids on natural convection in a porous medium. *Physica D, 40*(3), 393–402.

Shenoy, A. V. (1993a). Darcy-forchheimer natural, forced and mixed convection heat transfer in non-newtianian power-law fluid-saturated porous media. *Transport in Porous Media, 11*(3), 219–241.

Shenoy, A. V. (1993b). Forced convection heat transfer to an elastic fluid of constant viscosity flowing through a channel filled with a Brinkman-Darcy porous medium. *Warme- und Stoffubertragung, 28*(5), 295–297.

Wang, C., & Tu, C. (1989). Boundary layer flow and heat transfer of non-newtonian fluids in porous media. *International Journal of Heat and Fluid Flow, 10*(2), 160–165.

Wang, C., Tu, C., & Zhang, X. (1990). Mixed convection of non-newtonian fluids from a vertical plate embedded in a porous medium. *Acta Mechanica Sinica, 6*(3), 214–220.

Nomenclature

a	radius of rotating straight circular tube
$\overline{a}$	radius of curved tube, i.e., helical coil
a'	exponent of Grashof number in Eq. (4.40)
a''	exponent of Grashof number in Eq. (4.93) and defined by Eq. (4.94)
a''_0	constant in Eq. (6.5)
a_i	coefficient in the choosen velocity profile in Eq. (6.30)
A	coefficient in Eqs. (2.29) and (2.38) and defined in Eqs. (2.30) and (2.44)
A_1	dimensionless velocity term defined in Eq. (4.16)
A_c	cross-sectional area
A_i	coefficient in Eq. (2.39) and defined by Eq. (2.44)
A'	coefficient in Eq. (2.7) and Eq. (5.4) and defined in Eq. (2.8) and Eq. (5.5)
$\overline{A}$	constant in Eqs. (3.13) and (3.85) that is determined from the equation of continuity of secondary flow
b_0	function defined in Eqs. (5.40) and (5.41)
b_i	coefficient in the chosen temperature profile in Eq. (6.31)
b'	exponent of Prandtl number in Eq. (4.40)
b''	exponent of Prandtl number in Eq. (4.93) and defined by Eq. (4.95)
B	coefficient in Eqs. (2.29) and (2.38) and defined in Eq. (2.31)
B_1	coefficient in the profile for the dimensionless boundary-layer thickness in Eqs. (4.30) and (6.35)
B_2	coefficient in Eq. (4.31)
B_i	exponent in Eq. (2.39) and defined same as B in Eq. (2.31)
B'	exponent in Eq. (2.7) and defined in Eq. (2.9)
B''	coefficient in Eq. (5.4) and defined in Eq. (5.6)
$\overline{B}$	constant in Eqs. (3.13) and (3.85) that is determined from the equation of continuity of secondary flow
C	coefficient in Eq. (4.40)

A. Shenoy, *Rheology of Drag Reducing Fluids*,
https://doi.org/10.1007/978-3-030-40045-3

$C(\xi.\mathrm{De})$	correction function, defined by Eq. (2.15)
C_D	drag coefficient in Eq. (6.4)
C_E	coefficient equal to $\frac{\beta'' \epsilon^2}{\lambda''(1-\epsilon)}$ and dependent on the pore geometry
C_{fx}	local skin friction coefficient defined by Eq. (5.15)
C_p	specific heat per unit mass
C_0	function of β as defined by Eq. (4.66)
C_1	function of β as defined by Eqs. (4.28) and (4.67)
C_2	function of β as defined by Eqs. (4.29) and (4.68)
C' (1, De)	first derivative of correction function evaluated at the centerline, defined by Eq. (2.16)
C''	coefficient in Eq. (5.4) and defined in Eq. (5.7)
$\overline{C}$	pressure gradient in z-direction, $-\frac{\partial p^*}{\partial z}$ as given by Eqs. (3.5) and (3.95)
d_p	particle diameter (for irregular shaped particles, it is the characteristic length for average-size particle)
D	diameter of pipe (m)
D_c	dimensionless value of D_1 as given by Eq. (3.52)
D_1	characteristic angular velocity component in Eqs. (3.41), (3.111), and (3.136)
D_2	coefficient in Eqs. (3.67), (3.67), (3.67), (3.67), (3.67), and (3.67)
D_3	coefficient in Eqs. (3.67) and (3.136)
D_{ei}	equivalent diameter for inner pipe of the annulus defined by Eqs. (2.50)
D_{eo}	equivalent diameter for outer pipe of the annulus defined by Eqs. (2.51)
De	Deborah number defined by Eq. (2.10)
f	Fanning friction factor defined in Eqs. (5.19) and (5.20)
f'	reduced Fanning friction factor appearing in Eq. (5.57)
f_c	friction factor for curved tubes given by Eq. (3.66)
f_p	resistance or drag on a single particle defined by Eq. (6.4)
f_r	friction factor for rotating tubes given by Eq. (3.134)
f_1, f_2, f_3, f_4, f_5	functions defined by Eqs. (3.59), (3.60), (3.61), (3.62), and (3.63) for curved tubes and Eqs. (3.125), (3.126), (3.127), (3.128), and (3.129) for rotating straight circular tubes
F	roughness function defined by Eqs. (2.77) and (2.78)
F_i	function of geometry defined in Eq. (2.72)
F_o	function of geometry defined in Eq. (2.73)
F_N	roughness function for Newtonian fluids defined by Eq. (2.79)
F_R	total resistance to flow offered by N particles in porous media defined by Eq. (6.2)
$F(x)$	arbitrary function of x defined by Eq. (3.85)
g	acceleration due to gravity (m/s^2)
g_x	component of acceleration due to gravity in the x-direction defined by Eq. (4.45) and appearing elsewhere

G	constant in Eq. (2.13)
Gr	Grashof number defined in Eq. (5.53)
Gr_c	characteristic Grashof number for a drag reducing fluid defined by Eq. (4.13)
Gr_x	local Grashof number for a drag reducing fluid defined by Eqs. (4.19) and (4.74)
h	local heat transfer coefficient in Eq. (4.77)
h_w	coefficient of heat transfer at the wall defined by Eq. (4.8)
h_{av}	average heat transfer coefficient in Eq. (4.98)
i	integer associated with the coordinate system in Eq. (4.84)
I	function defined in Eq. (5.26)
I_v, I_t	functions associated with deviations from unity and defined in Eqs. (4.86) and (4.87)
j	integer associated with the body shape in Eq. (4.85)
k	thermal conductivity of the fluid
K'	intrinsic permeability of the porous media for flow of constant viscosity fluids defined by Eq. (6.12)
l_c	characteristic length
L	length of geometric shape (m) appearing in Eqs. (2.40) and (2.41), Eq. (4.98)
m	coefficient in the summation series given by Eq. (5.2)
$\overline{m}$	function defined by Eq. (5.30)
m_t	exponent associated with the wall temperature distribution as given by Eq. (4.88)
N	total number of particles defined by Eq. (6.3)
Nu	Nusselt number appearing in Eqs. (5.59) and (5.60)
Nu'	reduced Nusselt number appearing in Eqs. (5.59) and (5.60)
Nu_x	local Nusselt number for a drag reducing fluid defined by Eq. (4.38)
$Nu_{x,D,F}$	local Nusselt number for Darcy forced convection flow given by Eq. (6.45)
$Nu_{x,D,M}$	local Nusselt number for Darcy mixed convection flow given by Eq. (6.63)
$Nu_{x,D,N}$	local Nusselt number for Darcy natural convection flow given by Eq. (6.39)
Nu_{Lav}	average Nusselt number based on the characteristic length L and defined by Eq. (4.98)
p	pressure appearing in Eq. (3.78) and Eq. (6.1) and elsewhere
$p*$	pressure in rotating system defined by Eq. (3.78)
P	"P-function" defined by Eq. (5.13)
Pe_{cM}	characteristic Peclet number for mixed convection defined by Eq. (6.56)
$Pe_{x,D,F}$	local Peclet number for Darcy forced convection flow given by Eq. (6.46)
Pr	Prandtl number equal to $\mu C_p/k$

$\mathrm{Pr_c}$	characteristic Prandtl number for a drag reducing fluid defined by Eq. (4.13)
$\mathrm{Pr_x}$	local Prandtl number for a drag reducing fluid defined by Eq. (4.20)
ΔP	pressure drop ($\mathrm{N\ m^{-2}}$) appearing in Eqs. (2.40) and (2.41)
q	function of β defined by Eqs. (4.25) and (4.65)
q_w	heat flux at the wall defined by Eq. (4.9)
r	distance in the radial direction
$\bar{r}$	function of β as defined by Eq. (4.32)
$r*$	function representing geometric configuration in Eqs. (4.44a) and (4.44b)
r_i	radius of inner pipe of annulus (m)
r_m	radius of maximum velocity (m)
r_o	radius of outer pipe of annulus (m)
R	radius of circular straight pipe (m)
$\bar{R}$	radius of curvature for curved tubes (m)
R_p	particle radius in Eq. (6.5)
Ra_c	characteristic Rayleigh number defined by Eq. (6.21)
$\mathrm{Ra_{x,D,N}}$	local Rayleigh number for Darcy natural convection flow given by Eq. (6.40)
Re	conventional Reynolds number for constant viscosity fluids defined in Eq. (5.54) and Eq. (6.5)
Re'	reduced Reynolds number appearing in Eqs. (5.57) and (5.58)
$\mathrm{Re}*$	Reynolds number based on pipe radius and friction velocity
$\mathrm{Re_i}$	Reynolds number corresponding to the inner pipe as defined in Eq. (2.48)
$\mathrm{Re_o}$	Reynolds number corresponding to the outer pipe as defined in Eq. (2.49)
$\mathrm{Re_x}$	Reynolds number for drag reducing fluids defined by Eq. (5.27)
$\bar{s}$	function of β as defined by Eq. (4.33)
St	Stanton number defined by Eq. (5.38)
S_{tx}	Stanton number defined by Eq. (5.16)
T	temperature (°C or °K)
T_e	ambient temperature at edge of the boundary layer (°C or °K)
T_w	temperature of the wall (°C or °K)
T_∞	ambient temperature of bulk of the fluid (°C or °K)
ΔT	temperature difference (°C or °K) defined by Eq. (4.52) and appearing elsewhere
u	streamwise velocity component ($\mathrm{ms^{-1}}$) along the x-coordinate and in porous media it represents the Darcian or superficial velocity defined by Eq. (6.6)
u_e	axial velocity at edge of the boundary layer ($\mathrm{ms^{-1}}$)
u_m	maximum axial velocity at pipe centerline ($\mathrm{ms^{-1}}$)
u_s	seepage velocity (local average velocity of flow around the particle in Eq. (6.6))

u_w	wall slip velocity (ms^{-1}) in Eqs. (6.17) and (6.52)
u_1	dimensionless velocity component defined in Eqs. (4.16), (6.23), and (6.59)
u^+	dimensionless velocity defined by Eqs. (5.4), (5.48), and (5.49)
$u*$	friction velocity (ms^{-1})
u^*_{crit}	critical friction velocity demarking the onset of drag reduction (ms^{-1})
u^*_i, u^*_o	friction velocity (ms^{-1}) for inner and outer pipe of annular ducts, respectively, whose relationship is given by Eq. (2.43)
u^+	dimensionless axial velocity
$u^+{}_i, u^+{}_o$	dimensionless axial velocity for inner and outer pipe of annular ducts, respectively
U_c	characteristic velocity (ms^{-1}) defined by Eq. (4.15)
U_{cF}	characteristic velocity (ms^{-1}) for forced convection defined by Eq. (6.43)
U_{cM}	characteristic velocity (ms^{-1}) for mixed convection defined by Eq. (6.54)
U_{cN}	characteristic velocity (ms^{-1}) for natural convection defined by Eq. (6.19)
U_1	velocity scale near pipe wall (ms^{-1}) as given by Eqs. (3.44) and (3.114)
$\overline{U}$	average velocity (ms^{-1})
$\overline{U}_i, \overline{U}_o$	average velocity (ms^{-1}) for inner and outer pipe of annular ducts, respectively, defined by Eq. (2.52)
v	axial velocity component in the y-direction (ms^{-1})
v_m	mean axial velocity component for curved tube (ms^{-1})
w	velocity component in the z-direction (ms^{-1})
w_c	dimensionless value of w_1 for curved tube given by Eq. (3.53)
w_m	maximum velocity value of w at the center of the rotating straight tube (ms^{-1})
w_r	dimensionless value of w_1 for rotating straight tube given by Eq. (3.122)
w_1	axial velocity component at the edge of the boundary layer for curved and rotating straight tubes (ms^{-1})
We	Weissenberg number given in Eq. (6.5)
$We_{D,F}$	local Weissenberg number for Darcy forced convection flow given by Eq. (6.44)
$We_{D,N}$	local Weissenberg number for Darcy natural convection flow given by Eq. (6.26)
x	axial distance along the surface from the leading edge (m)
x_1	dimensionless distance defined in Eqs. (4.16), (6.23), and (6.59)
x_e	entrance length (m)
x_{ei}	entrance length (m) for inner pipe of the annulus given by Eq. (2.69)
x_{eo}	entrance length (m) for outer pipe of the annulus given by Eq. (2.70)

y	distance from wall normal to the surface (m)
y_1	dimensionless distance defined in Eqs. (4.16), (6.23), and (6.59)
y_{SL}	laminar sublayer thickness (m)
$y+$	dimensionless distance from wall
y_i^+, y_o^+	dimensionless distance from the wall for inner and outer pipe, respectively
z	direction in the coordinate system

Greek Letters

α	coefficient in Eqs. (2.25), (3.10), (4.4), and (5.20) whose values for varying De are given in Tables 2.1 and 2.2
$\alpha*$	integral defined by Eq. (3.133)
$\overline{\alpha}^*$	coefficient appearing in Eq. (2.5)
α''	is the shape factor in Eq. (6.4) which takes a value of $\pi/4$ for spheres
β	exponent in Eqs. (2.25), (3.10), (4.4), and (5.20) whose values for varying De are given in Tables 2.1 and 2.2
β_0	expansion coefficient of the fluid
β_1	pressure/stress gradient function defined in Eq. (5.3)
β''	is the shape factor in Eq. (6.3) which takes a value of $\pi/6$ for spheres
$\dot{\gamma}$	wall shear rate (s^{-1})
δ	thickness of boundary layer (m)
δ_1	dimensionless boundary-layer thickness defined by Eqs. (3.67), (3.136), (4.16), (6.23), and (6.59)
δ_2	coefficient in Eq. (3.136)
δ_c	dimensionless boundary-layer thickness for curved tubes defined by Eq. (3.51)
δ_i, δ_o	thickness of boundary layer (m) for inner and outer pipe of the annulus defined by Eqs. (2.67) and (2.68)
δ_r	dimensionless boundary-layer thickness for rotating straight circular tubes defined by Eq. (3.120)
δ_B	thickness of buoyant boundary layer (m)
δ_M	thickness of boundary sublayer plus buffer layer (m)
δ_M^+	dimensionless thickness of boundary sublayer plus buffer layer defined in Eq. (5.46)
η	dimensionless similarity variable defined by Eqs. (4.24) and (4.53)
ε	average height of the roughness projections in a rough pipe (m)
ϵ	porosity of the medium
ϵ	small quantity used for order of magnitude analysis to get Eqs. (3.29) and (3.102)
θ	angular coordinate
θ_1	dimensionless temperature difference defined in Eqs. (4.16), (4.52), (6.23), and (6.59)
θ_{fl}	fluid relaxation time (s)

κ	radius ratio r_i/r_o as given in Eq. (2.42)
κ'	von Karman constant appearing in Eq. (5.1) whose value is given in Eq. (5.10)
λ	radius ratio, r_m/r_o defined in Eq. (2.42)
λ''	coefficient in Eq. (6.5) dependent upon particle shape and size and which reaches a limiting minimum value of 3π for a single sphere
Λ_1	velocity component appearing in Eq. (4.5)
μ	fluid viscosity (Pa.s)
μ_b	density of bulk of the fluid (Pa.s)
μ_w	density of fluid at the wall (Pa.s)
υ	kinematic viscosity (m^2/s)
ξ	dimensionless distance from the wall, defined as y/R
$\overline{\xi}$	radial distance from the wall, defined as $= a - r$
$\hat{\xi}$	acceleration parameter defined by Eq. (4.54)
ξ_l	dimensionless viscous sublayer thickness appearing in Eq. (2.13) and defined as $y_{SL}^{+}2\sqrt{2/Re}\sqrt{f}$ using Eq. (2.23)
ρ	fluid density (kg/m^3)
$\overline{\rho}$	integrated fluid density (kg/m^3)
ρ_b	density of bulk of the fluid (kg/m^3)
ρ_w	density of fluid at the wall (kg/m^3)
$\sigma_1(De)$	function of De in Eq. (2.15), defined by Eq. (2.17)
$\sigma_2(De)$	function of De in Eq. (2.15), defined by Eq. (2.18)
τ_w	shear stress at wall ($N\ m^{-2}$) appearing in Eqs. (5.57) and (5.58)
τ'_w	reduced shear stress at wall ($N\ m^{-2}$) appearing in Eqs. (5.57) and (5.58)
τ_{rr}, $\tau_{\theta\theta}$	normal stress components ($N\ m^{-2}$)
$\tau_{r\theta}$, τ_{rz}	shear stress components ($N\ m^{-2}$) for the curved tube defined by Eqs. (3.117) and (3.118), respectively
$\tau_{r\theta}$, $\tau_{r\varnothing}$	shear stress components ($N\ m^{-2}$) for the rotating straight circular tube defined by Eqs. (3.48) and (3.49), respectively
τ_{wi}, τ_{wo}	shear stress at the wall ($N\ m^{-2}$) for inner and outer pipe, respectively
$\Delta\tau$	change in shear stress across buoyant layer ($N\ m^{-2}$)
ψ	stream function
ψ_1	function of β given by Eqs. (2.35) and (2.66)
$\overline{\psi}$	function of β given by Eqs. (2.28) and (5.35)
Ω	coefficient defined by Eqs. (4.60), (4.6), (4.60), and (5.22)
$\overline{\Omega}$	angular velocity

Author Index

A. Shenoy, *Rheology of Drag Reducing Fluids*,
https://doi.org/10.1007/978-3-030-40045-3

T

U

V

W

X

Z

Subject Index

A. Shenoy, *Rheology of Drag Reducing Fluids*,
https://doi.org/10.1007/978-3-030-40045-3

The manufacturer's authorised representative in the EU is Springer Nature Customer Service Centre GmbH, Europaplatz 3, 69115 Heidelberg, Germany. If you have any concerns regarding our products, please contact ProductSafety@springernature.com

Printed and bound by CPI Group (UK) Ltd, Croydon, CR0 4YY
05/10/2026
02228121-0010